BROADCAST TECHNOLOGY

BROADCAST TECHNOLOGY:
A Review

Dr. H.O. Srivastava

KALPAZ

Published by

Kalpaz Publications
C-30, Satyawati Nagar,
Delhi – 110052
Ph.: 011-47034999, 9811692060
www.kalpazpublications.com
E-mail: kalpaz@hotmail.com

Distribution Network
gyanbooks.com
India, USA, Canada, UK, Australia, France

ISBN: 978-81-21207-00-3 (HB)

Reprint in India, 2024
First Published, 2000

Printed at: Gyan Press, Delhi.

Broadcast Technology: A Review
Author: Dr. H.O. Srivastava

Dedicated to my lovely daughter
Aparajita

CONTENTS

PREFACE

Broadcasting is the most powerful medium of mass communication. However, the broadcasting has undergone a total transformation in the last 100 years. While the earliest transmitter used by Marconi was of spark type, full solid state digital transmitters are available today. Similarly, while Marconi used the Morse key to transmit his signal, the digital transmission of recorded speech and video using DAT/ DVT (Digital Audio/Video Tape) mixed with auxiliary data like paging /teletext is the norm of the day. In truth the new millennium media systems are going to take over the 'traditional' media completely. At present, the choice of content is limited to the station selected, the time of day, and the last resort of all, for the person who 'hears' nothing worth 'listening', to the 'off' switch. However, the technology available now offers the potential for a new range of listening options and uses of radio. A new mass communication technology revolution has started promising a potential global access to limitless information. The revolution is at the doorsteps of a major change, a change that possibly only comes once every 100 years. Time is ripe for a complete rethinking on the way we could or should disseminate information.

During early days of broadcasting, it was felt to have a magical power. The impact of television broadcasting was demonstrated by the live relay of President Franklin Roosevelt's address of New Years' world fair in 1928. When Mahatma Gandhi (Father of the nation) visited a broadcasting station to record a special message for refugees from Pakistan, he exclaimed, "This is a miraculous power. I see 'Shakti' the miraculous power of God". On another occasion of inauguration of SW Transmitter from Calcutta, the noble laureate Sh. Rabindranath Tagore exclaimed,

"Hark to Akashvani up-surging

From the below.

From Earth to Heaven,

distance conquered.

In waves of light

Flows the music of man's divining,

Fancy's flight"

Initially defined as a point to multi-point transmission system, the technology has brought the broadcasting to a doorstep where the initial definition has blurred and the future is moving towards an integrated interactive information delivery system merging broadcasting, telecommunication and computers. Broadcasting has developed during all these years, to a new height absorbing the advancements in technology, to enlighten, educate and entertain the masses. However, many new technologies are in the offing. Future TV transmission will bring digital pictures and sound all the way into our homes. Using digital compression several TV channels can be transmitted in the bandwidth of a single analogue channel, so allowing more to be received and with greater clarity of picture and sound. Progress has been made in research and development on the use of digital broadcasting technology to integrate a number of broadcasting services into a single digital broadcasting wave, thereby ensuring easy viewer access. Progress has also been made in research and development on ISDB (Integrated Services Digital Broadcasting) as a new broadcasting system to provide advanced multimedia services. Cable companies will use the technology to offer more channels. However, there is a price to pay: In general the greater the number of channels, the more the compression, the poorer the pictures. Services such as video on demand (VOD) may resort to heavy compression so that any selected movie can be on the cable within a tightly specified time frame but looking no better than current VHS...or maybe worse.

I would like to quote Bob Scott, ABC Radio. In his words, "The...future will be characterized by many competing means of delivery, for all types of electronic information, education and entertainment services. In all probability, the onward march of convergence will see a degree of integration of these means...demand critical reappraisal and change to existing radio...(and will)...see consumers using radio and associated services in new as well as traditional ways." Thus we are at the threshold of an exciting age and it is worth to have a look on broadcast technology. In this book, author tries to give a glimpse of broadcast technology.

Chapter 1 has been devoted to the historical development of broadcasting in the last 100 years. This chapter starts from the very definition of broadcasting, initial work of Guglielmo Marconi, earliest radio transmitters, development of FM and television transmitters, works of Nipkow, Baird, etc. development of camera and start of BBC Television. It further discusses development of the recording and the receipt systems. The broadcast regulations also evolved along with the development of broadcasting. The regulatory changes have been described in brief. The chapter finishes with few milestones, interesting events and anecdotes.

Chapter 2 is on Broadcast Basics. Starting from the broadcasting services including type of transmissions, it describes the basic working of TV system, video basics, various TV standards, lighting techniques, audio system, acoustics and radio frequency spectrum etc. Chapter 3 is devoted to Recording and Studio Techniques. It includes earliest recordings and studios, microphones and their characteristics, stereo, surround and quadraphonic sound, television cameras, audio recording and playback devices, video recording formats, mixing of audio and television film scanning. Chapter-4 describes production techniques. It covers broadcast processes, actors in the scene, video script, the camera shots, video editing etc.

Chapter 5 describes the basic transmission system including the different type of modulations, transmitter components, brief description of DAB, HDTV, Satellite transmission system, and propagation theory and transmission antennas. Chapter 6 describes the principle of broadcast receipt systems, new advancement in the receipt systems in view of the merger of broadcast, computer and telecommunication systems, and also the TV antenna connection for VHF/UHF reception. Chapter 7 briefly describes the test signals, charts and procedures.

Chapter 8 is devoted to broadcast network and its development in India. This is country specific detail and has been included for use as reference. Appendix 1 gives the colour and broadcast system by the country. Appendix II includes few interesting test cards used in different countries. Source, Credit and References follow these two appendices. This includes a number of Internet sites, which have got excellent materials, articles and books. The book has in the end a Glossary, which is a compendium of broadcast terms.

The book is intended to cater to all those persons who want to know as to what the broadcasting involves. This is also targeted to many broadcast professionals who work in one or other domain of broadcasting and want to know the subject as a whole. Similarly this book will work as an introduction to the broadcast subject for station directors /managers, who quickly want to grasp it.

ACKNOWLEDGEMENTS

The book would not have been possible without the assistance of my wife who had been a constant source of inspiration. I am amazed at all the information on this subject available on the World Wide Web. I have been guided by this information in writing this book. Because of importance of the source, I have given the references at the end of the chapters and also at the end of the book where the URLs have been mentioned. I acknowledge the help taken from many of the Internet sites.

1

HISTORY OF BROADCASTING

"RADIO BROADCASTING ADDED A TOTALLY NEW DIMENSION TO MODERN COMMUNICATION BY BRINGING THE OUTSIDE WORLD INTO THE INDIVIDUAL HOME. THE HISTORY OF RADIO, HOWEVER, WAS FAR MORE COMPLEX THAN THE HISTORY OF PREVIOUS MEDIA BREAKTHROUGHS. THE BROADCASTING SYSTEM TIED TOGETHER A BUNDLE OF TECHNOLOGICAL AND SCIENTIFIC THREADS THAT HAD BEEN DANGLING FOR A GENERATION."

DANIEL J. CZITROM, "THE ETHEREAL HEARTH: AMERICAN RADIO FROM WIRELESS THROUGH BROADCASTING," MEDIA AND THE AMERICAN MIND, CHAPEL HILL, NC: UNIV. OF NORTH CAROLINA PRESS, 1982. P. 60.

INTRODUCTION

The eternal words of Daniel J. Czitrom quoted above, are as true today as it were in early days of radio broadcasting. From its early theory, to a hobby for the 'tecl ies' of the early 1900s — radio and TV became first mass medium capable of disseminating information instantly from one to many. It has been a technology that has influenced growth of nations, and planet Earth. Radio, evolving ever as more affordable technology, has become a fundamental component of modern culture and has changed the world. While broadcast industry continues its transition to digital, this chapter presents some of the events, inventions and notable moments that made broadcasting what it is today.

In its exact sense, the term "broadcasting" includes only such signals as are sent out by means of radio waves for general reception.

A signal directed from one point to multipoint in code so that it is intelligible to masses, is certainly "broadcast" as the average individual or listener/viewer would define it. The listener's concept of "broadcasting" refers more specifically to speech or music sent out for the information, entertainment or education of the general public. Consequently, broadcasting as we know it today, may be said to have had its true birth in early 1907, when Lee de Forest constructed the first means of modulating an arc transmitter with voice impulses and began sound broadcasting from atop the 12 storied Terminal Building at 42nd St., Park Ave., in New York City[1].

No revolution has shaped modern mass media more profoundly than the recording revolution. To record is the ability to create an exact duplicate of an act of communication, to preserve and replicate this act quickly and cheaply, and distribute countless copies of this act to consumers everywhere, so that they can replay it over and over again whenever they so desire. Thomas Edison added the recordability of sounds in 1877 with his phonograph[2] and of motion in 1893 with his kinescope[3]. Although, in early experimental stage, the broadcasting was live, but soon recorded broadcasts were introduced due to it's inherent advantages of repeatability and flexibility of editing / dubbings.

Thus, the broadcasting quickly evolved into three distinct technological developments. The sound recording system at the studio end, the transmission system and the receipt system. Till a couple of year's back, broadcasting was essentially analog and used a variety of technologies in recording, transmitting and receiving signals. However, the first regular transmission of Digital Audio Broadcasting by British Broadcasting Corporation from September 1995[4] has brought into focus many advantages of digital broadcasting viz. high fidelity, digital editing, enhanced channel allocations using compression, storage techniques, etc.

The present era is of digital broadcasting with convergence of broadcasting, computers and telecommunication. The original definition of broadcasting as point to multi-point communication system is getting blurred with every passing day. Many point-to-point communications like paging has become possible using broadcast services and vice-versa. Many new type of services viz. Video-On-Demand[5], Interactive TV[6], WEB TV[7] etc. are in offing and broadcasting in the present form is likely to undergo a sea-change in future. The future will be characterized by many competing means of delivery, for all types of electronic information, education and entertainment services. In all probability, the onward march of convergence will see a degree of integration of these means, demand critical reappraisal, change the existing radio and will see consumers using radio and associated services in new as well as traditional ways. The countdown is now on for the new millennium.

A BRIEF HISTORY OF BROADCASTING

TRANSMISSION SYSTEM

Sound Broadcasting

The history of broadcasting dates back to December. 12, 1901, when Guglielmo Marconi made a successful experiment of transmitting the radio waves across the Atlantic. Marconi received the letter "S" in St. Johns, Newfoundland, by Morse code using a spark transmitter[8]. John A. Fleming was at the transmitter site in England Fig. 1.1. However this technological feat was possible due to the much earlier works of James Clerk Maxwell who formulated

Fig. 1.1. Marconi's antenna farm at Poldhu, England.

"Maxwell's Equations" for propagation of electromagnetic wave, in 1864 and Thomas A. Edison who recorded first sound -"Mary had a little lamb." on cylinders in 1877. Marconi established the world's first Trans-Atlantic commercial wireless service in 1907 with stations at Clifden, Ireland and Glace Bay, Nova Scotia.

The earliest radio transmitters including that of Marconi (1888-1905) employed spark technology. These transmitters used a spark between two electrodes to produce signals similar to the lightening that makes a radio crackle. Using a small-scale discharge of energy, spark transmitters produced crackles in headphones at distant receivers. Unlike natural lightening, however, the spark in the transmitter was turned on and off by a telegrapher's key, making it possible to communicate by code. Spark transmitters were relatively simple devices, and they could be built in any desired size, from lunchbox-sized portables to massive units that filled huge buildings. But even after years of development, they had many limitations. These transmitters covered a limited area, consumed huge power and only a few transmitters could be operated simultaneously in a particular area without interference. A thunderstorm could be heard all over the radio dial, therefore tuning was difficult. Marconi used rotary synchronous gaps in his spark transmitter.

Fessenden quickly recognized that continuous wave (CW) transmission was required for speech, and he could transmit and receive Morse code better by the continuous wave method than with spark transmitter that Marconi was using. He felt that if instead of sparking wild bursts of energy into the air, a way was found to concentrate that energy into a continuous wave of a single frequency, each station could be assigned its own frequency, allowing many more of them to operate without interfering each other. It would also be possible to cover greater distances with less power. But at the turn of the century, the only known method of producing continuous radio waves of any strength was with electromechanical devices known as alternators. In theory, if any alternator is operated fast enough, its output will be high enough in frequency enabling it to be used directly for radio. This lead to the development of alternator based radio transmitters. In 1906, the 'Alexanderson Alternator' was delivered to Fessendon's station. On Christmas eve in 1906, he broadcasted speech and music to surprised shipboard operators. The broadcast was done on 42 kHz using a technique called 'Amplitude Modulation'. The power of the transmitter was 1 kW. The programming included a female voice singing a Christmas carol, a violin solo by Fessenden, and an invitation to report on reception[9].

As CW systems were developed (1906-1912), Marconi sought to use his spark technology to achieve a semi-continuous timed spark that would approximate CW. Eventually Marconi's spark transmitter was replaced by the Fessenden/Alexanderson HF alternator.

Another historic development was taking place. In 1904, John Ambrose Fleming invented the first tube, the "Fleming Valve", or as he called it—an Oscillation Valve. His valve was a two-element rectifier, made by inserting a metal plate in one of Edison's electric light bulbs Fig. 1.2. Lee De Forest, in 1906 added a 'Grid' to the Fleming Valve, creating the 'Triode'. The three-element vacuum tube was well known by 1915 that was capable of regeneration and oscillation. It could therefore generate CW. One of the characteristics of the regenerative circuit was that as gain was increased beyond a certain point the circuit would begin to oscillate, or create a self-sustaining feedback. Armstrong noted this, and realized that the days of alternators and spark gap transmitters were over. The circuit used to

Fig.1. 2. Fleming Valve.

receive signals could also be used to send them. In 1913, Armstrong applied for a patent to use a vacuum tube as an oscillator. A station in Nauen, Germany began broadcasting on 16,900 meters or about 18 KHz (just above the range of hearing!) Another Station FL began broadcasting on 10,000 meters from the Eiffel Tower. Human voices were first broadcast in 1915 across the Atlantic Ocean, between Arlington, Virginia and the Eiffel Tower in Paris. Vacuum tube transmitters replaced the HF alternator.

World War I spurred transmitter-tube development. The War was over by 1919. The rise of CW followed in post-war years. Vacuum tube oscillators were replacing spark transmitters. Owen D. Young started the Radio Corporation of America (RCA) in 1920. The same year, Westinghouse built a 100 watt radio station named "KDKA", in a little shack atop its nine-storeyed factory in Pittsburgh Fig. 1.3. On November 2, 1920, Frank Conrad and Donald Little broadcasted election results from 8:00 pm till after midnight—an event that is credited with starting a rush to build stations, and purchase receivers. By late in the year, radio was being acclaimed as the newest form of entertainment for the home.

Fig. 1.3. KDKA Rooftop Shack in the early 1920's.

In 1920, Marconi established the first short-wave radio link between London and Birmingham, England on 20 MHz. Although most experimenters and pioneers used the longer waves, Marconi never did fully abandon his efforts to use the short wave bands. Describing 20 MHz as a 'short wave' is not really very accurate. One wavelength at 20 MHz is actually close to 50 feet in length. But in comparison to 833 KHz—one of the first frequencies assigned to broadcasting—it was short. 833 KHz has a wavelength of over 1, 120 feet. The local FM station on 99.5 MHz has a wavelength of about 10 feet, and it uses a microwave link to the transmitter on 950 MHz with a wavelength of just under one foot. These frequencies are still much longer in wavelength than many which, are being used for radio communications all around us now. By 1930, interest in short wave listening grew and short wave broadcasting started.

The Marconi Company started '2LO' in London—Britain's first broadcasting station in 1922. By 1924 spark was forbidden on the new 80, 40, 20 and 5 meter amateur bands. But spark was still used on the lower bands, and for another decade or more in the maritime service as a back-up for distress messages on the international distress frequency of 500 KHz (600 meter) right up to the beginning of the World War II. Had it not been for the War, spark would have been completely phased out in the maritime service except for emergency (lifeboat) purposes by the end of 1939.

A single radiator of an antenna radiates nearly equal amounts of RF energy in all directions. By placing longer and shorter elements —Reflectors and Directors—Near the drriven element, the signal can be made stronger in a specific direction. WSUN (AM) in Clearwater, Fla., installed first directional antenna in 1932.

Edwin Armstrong developed his theory to use FM in 1934 and demonstrated FM transmission in 1935. Armstrong operated W2XMN —a 50 KW FM station at Alpine, N.J. in 1939. By 1940, FM gained public interest as 'noise free', high fidelity broadcasting system. In the late 1950s and early 1960s the first test transmissions in stereo began by the BBC and then by RTE Radio in mid-1970s and finally by RTE Radio 2 (now 2FM) at the end of the 1970's. Thereafter, the FM band drew increasing numbers of listeners to popular as well as classical music, and when commercial FM stations began broadcasting in Ireland in 1989, they were allocated FM frequencies only. As a result, FM broadcasts now draw higher audience ratings than AM broadcasts. In India, FM transmitters are used for high quality classical and Indian music.

Television Transmission

Television is regarded by some as man's greatest invention, possibly more life sustaining than fire and certainly more entertaining than the wheel. There is lot of confusion about the question as to who invented television? People in the US credit Jenkins or Farnsworth with this invention. The Japanese believe it was Takayanagi. In Russia it was Boris Rosing. In France it was Belin and Barthelemy. In Eastern Europe, it was Von Mihaly and in Germany, Karolus. In the UK, the credit goes to Campbell-Swinton for concept and Baird for practical demonstration.

Paul Nipkow Fig. 1.4 had invented a method of mechanical scanning for television as early as in 1884. This was basically a disc with a single spiral of lenses or apertures on it Fig. 1.5. Each lens corresponded to a line of the television picture. One rotation would give one television frame. Not only was it simple to build (for a small number of lines), it could be used for both camera scanning and display scanning. The Nipkow disc was used by several TV pioneers as the basis for their television system.

Fig. 1.4. Paul Nipkow (1860-1940).

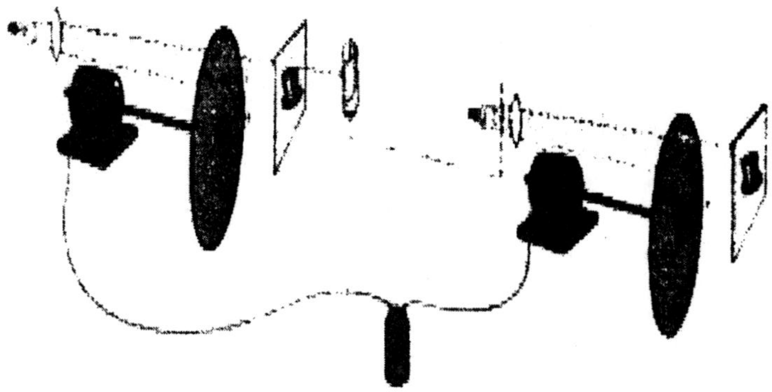

Fig. 1.5. Nipkow's System.

The Scottish inventor, John Logie Baird, succeeded in becoming the first to record television. Baird took Nipkow's scanning disc idea and the latest in electronics and developed this into the first demonstration of 'true' television in London in January 1926 Fig. 1.6. Recording the television signal in the 1920s was much simpler. With only 30 lines per picture (television frame), the highest frequency present was low enough to be audible. The video signal could therefore be recorded as an audio signal onto a disc. He transmitted the very first television picture in 1926 from one room to another. In 1927 he successfully sent a moving image along telephone wires from London to Glasgow, and the following year he achieved the first Trans-Atlantic television broadcast. His transmission of the image of a face across the Atlantic in 1928 was epoch breaking. The first experimental TV station began broadcasting in 1928—WGY in Schenectady, NY. Colour television, stereoscopic television and televisions by infrared light were all demonstrated by Baird before 1930[10,14].

Fig. 1.6. John Logie Baird and his television apparatus in 1926.

The BBC started broadcasting television on the Baird 30 line system experimentally from 1929. The first simultaneous sound and vision play was broadcast in 1930 followed by regular programmes of entertainment from 1932. Transmissions on the BBC continued until 11th September 1935 as developments in fully electronic television led to the demise of the 30-line service. The BBC replaced it with the world's first regular high-resolution (405 lines per picture) television service in 1936[15,16]. TV was first demonstrated at the New York World Fair in 1939. Colour TV was demonstrated for the first time in 1941. After World War II, television was relatively common in the USA and the UK. By 1949, both CBS and NBC had announced the development of colour television systems. CBS's system followed the path John Baird attempted and was known as the whirling disc system. The system placed the red, green and blue colours on successive fields. But when applied to the EIA standard, with only 60 fields per second, each colour would appear only 20 times a second, creating objectionable picture flicker. Eventually CBS increased the scanning rate to 180 fields per second and dropped the number of scanning lines from 525 to 405 lines. The cost of the CBS system was high. It required 43 per cent more channel space

(8.58 MHz) and would render all existing Black and White sets obsolete. NBC had taken a different approach and developed a colour television system that was compatible with existing B&W receivers. Regular colour television transmission began in USA, in 1950. NBC's Black and White compatible system was chosen by FCC on December 17, 1953 and remains a standard to this day.

> *"BBC Television was the world's first regular high definition television service which began on November the second, 1936, from Alexandra Palace, an old Victorian entertainment building in north London"*
>
> *Moss, Personal Interview.*

> *Source:* http://www.vaxxine.com/master-control/BBC/Bbc_tv.html

Development of Camera

Zworykin presented the first prototype iconoscope at a meeting of engineers in New York in 1929. The apparatus was built by RCA in 1933. It scanned the image in 120 lines, at 24 frames/second Fig. 1.7. and Fig. 1.8.

Fig. 1.7. The First Iconoscope.

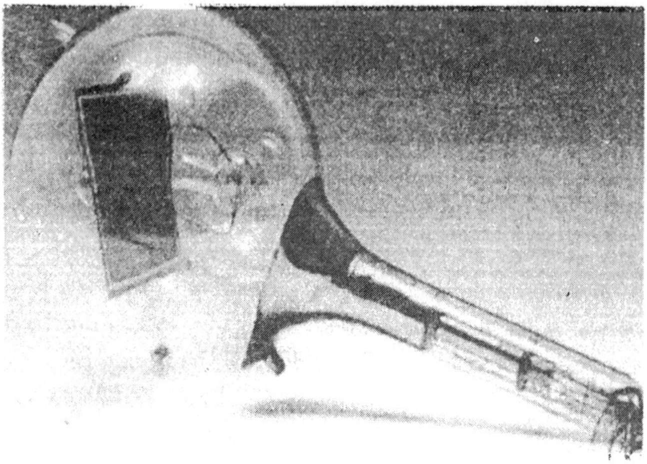

Fig. .1.8. The Iconoscope Glass Tube.

This first camera tube was more compact than the disc, much easier to use and more sensitive. The electron beam, which scanned the elements of the mosaic at a considerable speed, collected from each point all the photoelectric charge which, had accumulated there since the last visit. In the mechanical systems, the photoelectric cell received light from each point only during the very short period while it was actually being scanned.

COLOUR TELEVISION CAMERA DEVELOPMENT

The First Live Pick-Up Colour Cameras

RCA began live camera development for its colour television system in 1946-47. It initially used a flying-spot camera that could photograph an area, like the size of a person's face. A monochrome CRT was imaged through lenses onto the subject's face, and red, green, and blue sensitive photocells picked up the reflected light. Shortly thereafter, three monochrome image orthicon cameras onto an optical table were used. Light from the subject passed through a series of mirrors, which split the light into primary colours. The light was routed into three identical objective lenses of the three cameras. The camera therefore had only a single fixed focal length, and the three objectives were individually focused.

With increased interest by the industry in colour television, the Federal Communications Commission (FCC) on July 11, 1949 called for hearings to determine the feasibility of introducing colour service. Hearings began on September 26, 1949 and were, to last until May 26, 1950. Three competing methods of colour were proposed: the Dot Sequential approach of RCA; the Field Sequential method by CBS; and the Line Sequential proposed by Colour Television Incorporated. FCC rejected the Dot Sequential Colour Television System. The CBS field sequential colour system consisted of a rotating colour wheel of red, blue, and green filter segments in front of a monochrome camera, feeding a black and white CRT receiver viewed through a second rotating colour wheel. The two wheels were kept in phase synchronization, so that successive television fields could be viewed using identical colour primary filters to that of the camera. To overcome flicker, the field rate was increased from 60 to 144 fields per second resulting in 24 complete colour frames per second. To compensate for the increased field rate within the standard 6-MHz channel, the lines per frame were reduced from 525 to 405. A 29,160-Hz line rate resulted. This led to the basic incompatibility of the CBS system: no standard monochrome receiver could present an intelligible picture during the colour telecasts. FCC in its October 11, 1950 Report on Colour Television Issues, formally adopted the CBS system as the USA standard for colour television. On November 20, 1951, the National Production Authority prohibited the manufacturing of colour sets for general public sale. This order was the only control measure limiting an end product issued after World War II. Two years later CBS announced that it had no plans to resume its colour system.

This followed the development of TK 40 Fig. 1.9, TK 40A and TK 41 Fig. 1.10 colour cameras in 1954. These large, square front, cameras used a single objective lens. The cameraman could now select from four different focal lengths by rotating a lens turret. Director selection of cameras was done by switching the component R,G,B signals. A single colourplexer was used at the output of this switcher to encode the components of the selected camera into a composite NTSC signal.

The TK 41, introduced in late 1954, was the workhorse of the colour television industry for more than 15 years. The camera head and viewfinder, weighed 310 pounds, was 21 inches wide, 27 inches high, and, included lenses and viewing hood, which was nearly 5 feet in length. The camera and studio pedestal combination weighed 1000 pounds. A modern colour camera is shown in Fig. 1.11.

Fig. 1.9. TK 40 Colour Camera.

Fig. 1.10. TK 41 Colour Camera.

Fig. 1.11. A Modern Television Camera (1985).

Space Transmission

In 1942, British mathematician and science fiction writer Arthur C. Clark suggested using satellites to relay radio signals about 20 years before the first satellite, Sputnik was placed in orbit Fig. 1.12. Echo I reflected radio signals back to Earth in 1960. The concept turned into reality when satellite Telstar was used to transmit video images across the Atlantic in 1962. The first communication satellite was placed in geo-synchronous orbit in 1963. Astronauts sent the first live pictures and audio from the moon in 1969. Landsat I, the "eye-in-the-sky" satellite, was launched in 1972. By 1978, PBS started delivering programming by satellite.

In 1980, Intelsat V started relaying two colour TV channels on regular basis. In 1984, Ku-Band satellites were used for news feeds to broadcast stations. In U.S. TV networks began satellite distribution to affiliates in 1985.

Fig. 1.12. Arthur C. Clark's Concept Satellite.

Gentner introduced the first 'dial-up' remote control system for a transmitter in 1986. Same year, technology was developed to scramble the signals. This started the concept of pay and virtual channels.

DIGITAL SYSTEMS

In mid-1970s, BBC research engineers investigated the possibility of providing a digital radio service based on a NICAM stereo signal. They established a system with a small transmitter and a reception vehicle which, could receive and decode the signals using both a directional antenna, and a standard whip aerial. They found that the reception on the directional antenna was generally very good, but that the vehicle's rooftop antenna was not satisfactory over a significant proportion of the service area. The problem was generally caused by multipath reception which, resulted in considerable interference to the digital signal. The multipath reception is the receipt of two or more signals from different directions due to reflections. In a NICAM system interference starts to become a problem once the delay between the main and reflected signals becomes greater than about a microsecond. That corresponds to the reflected signal travelling about 300 meters further than the main signal.

It was concluded that such 'simple' digital systems, such as NICAM, could only be used when the signal is to be received on fixed directional antennas. This is particularly true in built-up or mountainous areas where reflected signals are more common.

The development of digital technology and compression techniques started the race for digital broadcasting. The need for CD quality audio and the mandate to make these signals immune to multipath interference lead to the development of Digital Audio Broadcasting (DAB) system. In 1992, 'USA Digital Radio' conducted its first 'on-air' tests of "in-band" on channel AM in Cincinnati, USA. However, the lack of an approved standard greatly hindered the development of this mode of broadcasting. In 1993, FCC selected the Motorola C-QUAM system of AM stereo transmission as the US standard—ten years after AM stereo was first authorized. BBC with other European partners started working on the DAB standard. To overcome the multipath interference, the signal was split into a number of separate parts and transmitted on a group of frequencies, spreading the sound data equally over a number of channels. A modulation system known as Coded Orthogonal Frequency Division

Multiplex (COFDM) was used where a large number of low data-rate carriers are packed as close together in frequency as is theoretically possible. As the bit-rate of each of the data carriers is very low, they are fairly insensitive to multipath interference.

Denmark began on-air testing of the Eureka 147 method of DAB on 237 MHz in 1994. The BBC—began introductory DAB service in 1995. DAB was started in Sweden in 1996, from four transmitters utilizing Eureka 147[17]. All India Radio started an experimental DAB service in Delhi in 1996.

The EBS—Emergency Broadcasting System—was developed in 1963. This was replaced by Emergency Alert System on January 1st, 1997.

RECORDING SYSTEM

The live production of radio shows was a deeply established tradition by ASCAP (American Society of Composers, Authors and Publishers) until 1907 when the first 'broadcast' of recorded programme was made, so the operator didn't have to talk. A sound recording is defined as a disc, tape, film, cylinder, or wire on which sound vibrations have been registered so that the sound may be reproduced as and when desired. In 1917, E.C. Wente at Bell Labs developed the condenser microphone to translate sound waves into electrical waves that could be amplified by the vacuum tube amplifier. In April, 1931, Leopold Stokowski invited Bell Labs to begin sound recording experiments with his Philadelphia Orchestra. Initial recorded broadcasts by NBC failed to achieve the high quality of reproduction. After a series of disappointing broadcasts of recorded orchestra in 1930-31, Bell Labs set up a test room at the Academy of Music in Philadelphia. Arthur C. Keller installed a vertical-cut recorder equipped with a new moving coil pickup with sapphire stylus that extended the dynamic range to 10,000 cycles. Coating the wax master with gold film and a layer of electroplated copper reduced surface noise. The duplicate release copies were pressed on cellulose acetate rather than shellac. In December, the first electrical recordings were made and continued throughout the 1931-32 concert season[18].

In 1946, Bing Crosby, a performing star, wanted to shift from live performance to recorded transcriptions for his weekly radio show on NBC sponsored by Kraft. But NBC refused to allow recorded radio programs (except for advertisements). The 16-inch aluminum program discs were made from master discs running at 78 rpm and holding only 4 minutes per side. The short playing time of 78 revolutions per minute (RPM) records meant that a 50-minute symphony would be packaged in the brown paper sleeves of a five-disc leather-bound "album" resembling the cover of photograph scrapbooks. Perhaps out of habit, people later referred to single disc LPs as albums. These records also presented editing and timing problems that often caused gaps or glitches in the flow of the 60-minute program. Also, the surface coating of the aluminum discs was still the same wax that Edison had used at the turn of the century, with the same limited dynamic range and frequency response.

In June of 1947, Murdo MacKenzie of Crosby Enterprises saw a demonstration of the German Magnetophone that Jack Mullin had brought back from Radio Frankfurt, along with 50 reels of tape, at the end of the War. This machine was one of the magnetic tape recorders that BASF and AEG had built in Germany starting in 1935. A half inch ferric-coated tape could record 20 minutes of high-quality sound per reel. However, the records weared out after 75 plays. Alexander M. Poniatoff, who had seen Mullin's demonstration of the machine one year earlier in San Francisco, ordered his Ampex Company (founded in 1944 from his initials A.M.P. plus the starting letters of "Excellence") to manufacture an improved version of the Magnetophone.

The decade following the end of World War II witnessed a "revolution in sound" recording. Crosby's Decca Company introduced FFRR (Full Frequency Range Recording) 78pm records with reduced size of the groove and a delicate elliptical stylus to track on the sides of the groove. The FFRR had the finest frequency response (80-15,000 Hz) of any recording process till the advent of magnetic tape recording. Columbia used Decca's methodology in 1948 on the new 33-1/3 rpm LP vinyl record. LP records used jeweled stylus (not steel) tracking at only 10 grams (not 200)[19]. Columbia Records' (owned by CBS) Peter Goldmark finally perfected long-playing (LP) records that could play at 33 1/3 rpm for over twenty minutes on each side in 1947.The LP or Long Play phonograph was introduced in 1948. The 45 rpm record was introduced in 1949, by RCA.

The late 1940s also saw magnetic tape recording emerge as a viable method of sound reproduction. The most important pioneering experiments in magnetic tape had been conducted in Germany starting in the late 1920s. By the mid-1930s a "magnetophone" that recorded and played on magnetic tape had been perfected. The Ampex Company joined by Magnecord, Webcor, Revere, and Fairchild, sold one million tape recorders to a rapidly growing consumer audio component market by 1953. The 1949 Magnecord tape recorder had stereo capability eight years before any vinyl record had it. Stereo Records reached the marketplace in 1958.

Mullin demonstrated the first video tape recorder on December 30, 1952, but the picture was blurry. He was also not able to solve the problem of high tape speed. Magnetic tape was demonstrated for recording television pictures by RCA in 1954. However, it was the Ampex team led by Charles Ginsburg that made the first videotape recorder. Rather than speeding tape across fixed heads at 30 mph, Ginsburg used rotating heads to record at a slant on tape moving at only 15 ips. The helical scan model VR-1000 was demonstrated at the NAB show in Chicago on April 14, 1956, and was an immediate success. By 1957 most TV production was done on videotape. Ampex developed a colour videotape system in 1958.

The Tape Cart (soon to become the 8 track tape) was introduced in 1960. Cassette tape was introduced in 1963. Philips[20] marketed the first encased audio tapes in 1964, and within a few years eight-track cartridges emerged as the front runner in a market that included four-tracks and cassettes. Dolby noise reduction was introduced in 1969[21]

In 1973, a company now known as Quantel (Quantised Television), developed the first practical analogue-to-digital converter for television applications. Now, more than 25 years on, digital technology is used throughout the television industry. Compact discs (CDs), on which master recordings are converted to digital information, providing clear, superior resolution, were introduced in 1980[22]. As videotape recording reached its stride in the 1980s, Sony combined the helical-scan recording technology used for video signals with the latest digital technology in audio recording. The result was the introduction in 1987 of Digital AudioTapes (DAT) and DAT recorders. By 1992, 80 per cent of recording studios in a Pro Sound News survey said they made use of a DAT machine. The Digital Compact Cassette (DCC) was introduced in 1992. Digital revolution continues the development of exciting and innovative new products and technologies with the aim to help users achieve better results faster and without constraints. The digital innovation has started a process that has fundamentally changed both the look of television and the way it is produced.

RECEIPT SYSTEM

Karl Ferdinand Braun introduced the use of a crystal detector as part of a wireless receiver in 1901 Fig. 1.13. Crystal receivers worked well for voice reception, since a crystal detector and a capacitor could remove the carrier from amplitude modulated signal, leaving only the audio wave that had ridden on the peaks of the carrier. This audio signal was converted back to sound by the headphone[25]. Reginald Fessenden invented the 'Electrolytic Detector' in 1902. Marconi invented the magnetic detector, in the same year.

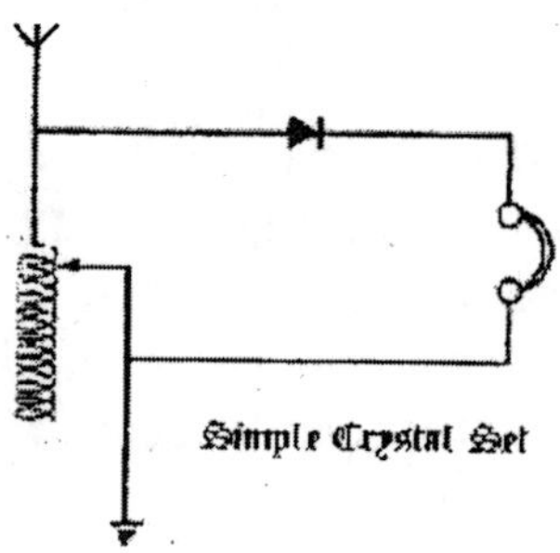

Fig. 1.13. A Receiver using Crystal Detector.

Fessenden invented 'Heterodyne reception' in 1904. In a heterodyne circuit, two frequencies are mixed to create a sum and difference of the original two frequencies. If a 100 KHz frequency and 60 KHz frequencies are mixed—two new frequencies of 160 KHz and 40 KHz can be obtained. Each of the new frequencies has the characteristics of the original two frequencies. In 1907, G.W. Pickard perfected the crystal detector and obtained a patent for the use of silicon in detectors. By 1911, young radio amateurs started building receivers.

Many engineers had struggled with the problems encountered in amplifying radio frequency signals above 200 KHz. Armstrong recognized that using a heterodyne method of reducing the frequency of the carrier would solve that problem. A heterodyne circuit mixes two frequencies to obtain a sum and difference frequency with the modulation intact. Armstrong announced the first superheterodyne circuit in 1920. Armstrong's circuit mixed the radio frequency with a local oscillator frequency to obtain a lower frequency—nearly 45 KHz, which was easy to amplify. Westinghouse purchased the Superhet patent from Armstrong, along with several patents from Reginald Fessenden and Michael Pupin.

Battery operated sets needed several different voltages to operate. The 'A' battery was the filament battery—providing low voltage (3 to 6 volts were common) for the tube filaments. These batteries would last from one to three weeks, and could usually be recharged. The 'B' battery provided plate voltage for the tubes (60 to 130 volts being common) and would last from two to six months. These could not be recharged. The 'C' battery provided bias or negative voltage to the tubes, and added to the longevity of the 'B' battery. The 'C' battery had to be replaced every six to twelve months. Radio experimenters spent over two million dollars for radio parts in 1920. Speakers began to replace headphones for listening in 1921. In 1922, Edwin Armstrong invented the 'Super-Regenerative' receiver. In 1923, Edwin Armstrong invented the first 'Portable' radio—a wedding gift to his wife. 'Radio fever' in early 1923 took a noticeable leap. The lyric of a popular (English) dance tune reflected the rising mood of enthusiasm:

"If you haven't got a wireless set
And fixed your mast upright,
You're losing half the fun we get
By listening in at night.
The world is mad about the game
Why don't you start and do the same."

Source: Briggs, Susan, Those Radio Times

By 1924, radio's—superhets, reflex sets, TRF's, and heterodynes were in use. In 1928, diode detectors received consideration by radio designers. By 1929, a typical AC TRF receiver employed type 226 tubes in the RF and AF amps, a 227 as a detector, type 71A for the output and a type 80 in the power supply. By 1930, the TRF, Tuned Radio Frequency receiver was still the leader, but many superhet receivers were being made. AVC or Automatic Volume Control was introduced in 1932.

'All-Wave' receivers were a hit in the year 1934, bringing in signals from foreign broadcasters. Most radios sold in 1936 employed an AFC circuit—Automatic Frequency Control. Cathode Ray tuning eyes (the Magic Eye Tube), Slide Rule tuning and sleek veneered cabinets were the big features of the receiver of 1937.

Several manufacturers sold the first television sets in 1939. The start of the European war in 1939, renewed interest in short-wave receivers. With the invention of transistor by John Bardeen, Walter Brattain, and William Shokley by 1947, radios started shrinking in size using miniature tubes, rectifiers, transformers, and printed circuit boards. Japanese upstart company—Sony marketed the first transistor radio in 1957 and the first transistor TV in 1960). In an effort to produce colour pictures on black and white sets, in 1962, several manufacturers reportedly tinted the CRT blue on top, and green on the bottom. Portable AM receivers lead in production—over five and half million were made in 1962.

The manufacture of portable AM/FM or FM sets grew at 750 per cent between 1960 and 1965. New vacuum tube receivers became almost non-existent by 1965 as transistors had taken over the market. Between 1922 and 1965, over 300 million radio receivers were manufactured in the United States alone. The year 1984 witnessed a television set that could be worn on the wrist. In 1985, Sony built a radio of the size of a credit card.

REGULATIONS

US Congress passed the Radio Act of 1912 to prevent amateur radio operators from interfering with government stations. In 1914, war broke out in Europe, and amateur licenses were suspended in almost all-foreign countries. In 1917, America entered the First World War, and all patent protection was set aside for the duration. Amateur radio experimenters pulled down their antennas and packed away their equipment by Government order. In 1922, two frequencies viz. 833 and 619 KHz, were authorized for broadcasting. A third frequency —750 KHz, was added later in the year. In 1924, the present A M band—550 to 1550 kilocycles, was assigned. A U.S. court decided in 1926 that the Secretary of Commerce had no power to regulate broadcasting—but could only issue licenses, resulting in a complete chaos on the broadcast bands as stations increased power to drown out the competition. The FRC (Federal Radio Commission) began to regulate broadcasting in 1927. Their first act was to revoke all licenses, and then assign frequencies and power levels. The Federal Communications Commission (FCC) was Created in 1934 by congress through the "Communications Act". Broadcasting standards for TV were announced in 1938, paving the way for commercial television stations. In 1941, FM broadcasting was authorized on 42-50 MHz. All amateur radio communication was halted by the war. In 1945, the FM band was changed from near 50 megacycles to the present 88 to 108 megacycles. This rendered many sets obsolete, and was a setback to Armstrong's development of FM as an alternative to AM[24]. The FCC announced three months freeze on new TV station applications in 1948. It lasted nearly four years when TV picked up steam. The FCC's three month' freeze on new TV station applications, imposed in 1948, was finally lifted in 1952. In India, after independence, the radio broadcasts were regulated by

Indian Telegraph Act 1985. FM Stereo Broadcasting was authorized in 1961. The FCC ruled in 1964 that AM-FM combo's must run different programs on the two stations at least half the time.

In 1972, new FCC rule permitted community access television. "Open Skies" allowed any U.S. firm to have communication satellites. AM Stereo was first authorized in the United States in 1982. Five competing systems struggled for dominance—as the FCC refused to settle on a standard. FCC allowed 'Short Spacing' of FM stations in 1988. FCC approved expansion of the AM broadcast band from 1605 to 1705 KHz in 1991. New station ownership rules were put into effect—a single group could own up to 18 AM and 18 FM broadcast stations. In 1993, the U.S. Congress instructed the FCC to begin collecting "Regulatory Fee" from broadcast stations. The first allotment plan for the expanded AM band was released in 1994. In 1995, the FCC revised rules to allow unattended operation of broadcast stations. The first broadcast station—WJDM - Elizabeth NJ at 1660 KHz—started operation in the expanded band in 1995. The FCC allocated spectrum for a new satellite delivered Digital Audio Radio Service (DARS) in the 2310 to 2360 MHz band. FCC adopted a standard for High Definition T.V. (HDTV) in 1996. The U.S. Congress mandated that the FCC collects over 152 million dollars in "Regulatory Fee's" from broadcasters in 1997. The FCC issued two licenses for DARS in 1997. In 1995, a judgment by Supreme Court of India pronounced the airwaves as the free property of any citizen.

GROWTH OF BROADCASTING: A FEW MILESTONES

It is worthwhile to see few important milestones in the growth of broadcasting. The chronology of the events for sound recording and the broadcast history are given in Table 1.1. A brief history of video and television technology is given in Table 1.2. The growth of broadcast stations and receivers in early years are given in Table 1.3. During the 98 years of the journey, the "radio" had few interesting uses and episodes. These are given in Table 1.4. A few landmarks in the process of development of digital TV is given in Table 1.5.

Table 1.1 : Brief History of Sound Recording and Broadcasting Technology

1864	James Maxwell's electromagnetic wave theory for radio wave propagation.
1877	First description of recording sound onto a cylinder or disc described by Charles Cros in France and Thomas Alva Edison in the United States.
1878	Thomas Edison patents the recording of sound onto discs and cylinders.
1887	Heinrich Hertz transmits and receives radio waves over short distances.
1888	Emile Berliner shows first example of a working "phonograph" playback device.
1888	Basics of magnetic recording put forth by Oberlin Smith.
1889	Danish inventor Valdemar Poulsen patents the first magnetic recorder.
1895	Development of first wireless telegraph system by Guglielmo Marconi.
1905	First electron tube developed by Sir Ambrose Fleming.
1906	First wireless communication of human speech.
1919	KDKA in Pittsburgh, Pennsylvania is licensed as the first broadcast radio station.
1925	First electronic recordings made with the use of a microphone.
1931	First experimental stereo recordings made by Bell Telephone Laboratories.
1933	Theory of frequency modulation (FM) for radio broadcasts by Edwin Armstrong.
1948	Introduction of Long Play (LP) record by CBS.
1948	First transistor introduced.
1954	Introduction of stereo tapes to the public.

1962	First stereo FM radio broadcasts.
1975 - 1978	Early digital recording made.
1980	Sony introduces the "Walkman".
1983	First CD player made available by Sony and Philips.
1995	Digital Broadcasting starts.

Table 1.2: Brief History of Film, Video, and Television Technology

1897	Development of the cathode ray tube by Ferdinand Braun.
1907	Use of cathode ray tube to produce television images.
1923	Patent for the iconoscope, the forerunner of the modern television picture tube.
Early 1930s -	RCA conducts black and white broadcasting experiments.
1936	First television broadcast made available in London.
1938	Initial proposal for colour television broadcast made by George Valensi.
1949	System developed to transmit chrominance and luminance signals in a single channel.
1950s	3D and Cinemascope formats introduced by Hollywood.
1954	NTSC standard for colour television broadcast introduced in the United States.
1966	PAL standard introduced in Europe.
1975	Sony markets the first Betamax VCR for home viewing and recording of video.
1976	JVC introduces the VHS format to the VCR arena.
1976	Dolby Laboratories introduces Dolby Stereo for movies.
1978	Philips markets the first video laser disc player.
1984	The first Hi-Fi VCR is introduced.

1985	The broadcast of stereo television.
1994	Standard agreed upon for high definition television (HDTV) transmission.

Table 1.3: Milestones in growth of broadcasting in early years

Year	*Broadcasting Stations*	*Receivers*	*Remark*
Dec. 1921	5		"I can hear it with the headphones lying on the table".
Sept. 1922	537	100,000[1]	
1923		500,000[1]	
1924	1400	3 million[2]	
1936		8 million[3]	3 out of 4 families have a radio in home.
1941	30 FM	13 million[1]	
1944		57 million[2]	Over 30 million U.S. homes.
1946		15 million[1]	Half a million FM receivers.
1947		800,000 FM[1]	
1949	2021 + 98 TV	4 million TV[2]	
1950		90 million radio[2]	Average 2 radios for every home in US.
		10 million TV[2]	
1951	108 TV		
1953	326 TV		

1955	7 million[1]	
1961	11 million[1]	
1995	585 million radio[2]	Average 5.6 radio in every U.S. home

1 receivers produced during the year

2 receivers in use

3 sets sold during the year.

Table 1.4. A Few interesting and notable events

Year	*Event*
1923	U.S. President Harding had a radio installed at the White House. Edwin Armstrong invented the first 'Portable' radio—a wedding gift to his wife.
1924	U.S. President Coolidge's cat is lost and found with the help of Radio.
1926	"Pay Radio" made its debut in Philadelphia. Placed in retail stores, these 5 ft. high units charge a nickel for 5 minutes of listening.
1937	The dirigible, Hindenburg crashed in flames at Lakehurst, N.J. on May 6th, 1937 and the tragedy was captured in an incredible live radio broadcast.
1938	Howard Hughes flew around the world and kept in touch by radio. The power of radio was demonstrated by Orson Wells, and the "Mercury Theater of the Air" - Panic is reported to be widespread as people believe the earth has been invaded by "Martians", and a "War of the Worlds" had erupted.
1941	The Japanese attacked Pearl Harbor. FDR spoke to the

	nation. That evening, Eleanor used radio to speak to the women of the nation.
1944	Germany made use of shortwave radio for propaganda broadcasts.
1946	Lee De Forest grew increasingly unhappy with the state of broadcasting. To him, radio had become 'a stench in the nostril of the gods of the ionosphere.' He addressed the National Association of Broadcasters at their annual meeting in Chicago.
1973	September 19th 1973, a 'Pirate Radio' station began to broadcast from a ship anchored some 3 miles offshore of Cape May, N.J. It was shut down the same day by the FCC.

Table 1.5: Landmarks in the process of development of digital TV

1975	Quantel demonstrated digital framestore synchroniser - DFS 3000.
1980	Digital still storage for on- air presentation developed.
1981	Paintbox is launched.
1982	Mirage is launched. It introduces the TV page turn and is the first digital effects machine able to manipulate 3D images in 3D space.
1986	Multi-layering becomes a practical proposition.
1990	The Picturebox stills store system is launched.
1994	Diskbased broadcasting started.

CONCLUSION

In this chapter, the history of broadcasting, few milestones and notable movements have been described.

REFERENCES

1. "An Overview of Broadcasting through 1940", at http://www.antique-radio.org/overview .html
2. http://sln.fi.edu/franklin/inventor/edison.html
3. http://lcweb2.loc.gov/paper/mpixhome.html
4. D. Cope, "Digital Audio Broadcasting", International Broadcast Engineer, 1996, pp. 28-30.
5. W. Hodge, et. Al., "Video-On-Demand: Architecture, Systems and Applications", **SMPTE Journal**, pp. 791-799, Sept. 1993.
6. C. J. Birkmaier, "A Commentary on Requirements for the Inter-operation of Advanced Television with the National Information Infrastructure", at http://www.eeel.nist.gov/advnii/birkmaier.html
7. B. Manasco, "Open Architecture DBS and the era of Customized Television", Via Satellite, Vol. 11, No. 3, pp. 467-471, 1995.
8. The First Sparks: The Birth of Radio at http://www.cinemedia.com.au/SFCV-RMIT-Annex/rnaughton/phd4410.html
9. Carter, Alden R., "Radio: From Marconi to The Space Age" at http://www. cinemedia.net/SFCV-RMIT-Annex/mavighton/MARCONI-1.html
10. 'Phonovision', "The World's First Recordings of Television" at http://members.aol.com/mcleandon/tv_1strx.htm
11. D. F. McLean, "Computer-based analysis and restoration of Baird 30-line television recordings", *Journal of the Royal Television Society*, Vol. 22/2 April 1985 pp. 87-94.
12. D F McLean, "The Recovery of Phonovision", Third International Conference on Image Processing and its Applications, *IEE Conf Publ 307*, July 1989, pp. 300-304.
13. Dinsdale, "Television", 2nd edn. 1928.
14. Moseley and Barton-Chapple, "Television Today and Tomorrow", 2nd edn. 1931.
15. T H Bridgewater OBE, "Just a Few Lines", Supplement to *Bulletin of British Vintage Wireless Society*, August 1992.

16. Bruce Norman, "Here's Looking at You", BBC and Royal Television Society, 1984.

17. "Eureka 147/DAB system Description", European Broadcasting Union Report No. DT/05959/III/C, December, 1992.

18. "Audio Recording : History and Development" at "http:// spot.colorado.edu/~rossk/ history/radio/ audiohd.html

19. http://www.music.sony.com/Music/Musicindex.html

20. http://www.philips.com/

21. http://www.dolby.com/

22. http://www.umn.edu/nlhome/g496/eric0139/papers/paper.html

23. "Just - 'The Cat's Whisker' : Magic Crystals" at http:// www.cinemedia.com.au/SFCV-RMIT-Annex/rnaughton/ phd4420.html

24. http://www.northernnet.com/bchris/

2
BROADCAST BASICS

BROADCASTING SERVICE

The broadcasting service is used to serve both international and domestic audiences. International audiences are served by *external service*, which are mostly short wave and medium wave radio transmitters. Domestic audiences are served by *home service* over-the-air radio and television systems. This service is also the major source of local news, sports, public affairs, etc.

EXTERNAL SERVICE

External service by its very nature requires the generation of signals that are intended to be transmitted across international borders. Consequently, transmission of these signals is subject to the ITU (International Telecommunication Union) Radio Regulations. For decades, governments have made increasing use of the electromagnetic spectrum to conduct public diplomacy by broadcasting speech and music throughout the world. The *external service* stations are administered by various medium and short wave frequencies depending on the time of day and season of the year. These broadcasts are made directly to receivers which are used by individuals throughout the world.

The analog high frequency transmissions suffer from distortion and fading in the ionosphere, from congestion in the bands that deliver an audio quality of medium frequency. Against this background, international broadcasters are examining the potential of delivering digital radio programmes in future to audiences worldwide as a direct to home (DTH) service using satellite. A consortium for DRW (Digital Radio WorldWide) set up in 1994 is addressing this issue[1].

HOME SERVICE

The domestic audiences are served by AM, FM, and TV broadcast stations employing analog radio transmissions designed for direct reception by home receivers.

AM BROADCAST STATIONS

AM broadcast stations operate on a channel in the 531-1602 KHz AM broadcast band. This band consists of 120 carrier frequencies beginning at 531 KHz and progresses in 9 KHz steps to 1602 KHz. The modulation of the radio carrier wave is amplitude modulation; hence, the AM reference. The operating power ranges from 1 kilowatts to 1000 kilowatts. Propagation in the AM broadcast band involves both the ground wave and skywave modes. The ground wave and skywave modes of AM broadcasting stations serve local and distant audiences, respectively. A disadvantage in AM broadcasting is its limited audio fidelity, relative to FM.

FM BROADCAST STATIONS

FM broadcast stations are authorized for operation on 80 allocated channels, each 180 KHz wide, extending consecutively from 100 MHz to107.9 MHz with standard deviation of 75 KHz. Better audio fidelity is a distinct advantage of FM radio over AM radio broadcasting. However, FM radio does not normally have the extensive service coverage areas that AM radio broadcasting enjoys.

TELEVISION BROADCAST STATIONS

Television is the world's most powerful form of communication. Every day it reaches out to millions of people to entertain and inform them with 'real life' images of the world around them. Commercial and educational television broadcast stations comprise the broadcast television organization. These stations operate on 6 MHz wide channels in the VHF and UHF frequency bands. The spectrum occupied by television broadcast comprises 72 MHz in the VHF band and 336 MHz in the UHF band.

Television signals can be transmitted using terrestrial ground transmission, cable or satellites. Terrestrial television uses radio and microwave techniques to transmit a composite signal of amplitude modulated (AM) pictures and frequency modulated (FM) audio via an antenna which covers a geographical limited area by line-of-sight. A receiving antenna enables the decoded picture signal to be displayed on a television screen and the audio signal through a loudspeaker.

Cable is another means of transmitting television. Cable is a local distribution mechanism that transmits the video signal via a fibre optic or copper wire cable directly to television sets. Cable television can link into national television networks by receiving terrestrial or satellite transmission, and relaying these via cable.

Satellite television makes use of a satellite in a stationary orbit around the Earth that acts like a giant mirror. The video signal is beamed up to the satellite by a ground station and then beamed down again over a large geographical area (known as the 'footprint' of the satellite). The parabolic receiving dish of the user receives the satellite signal. Satellite transmission is limited by the number of satellites available and by available frequencies for transmission. Satellites have to broadcast at frequencies that do not interfere with either terrestrial broadcast frequencies or other satellite broadcasts. As with terrestrial broadcasts, there is therefore limited bandwidth available for satellite television transmission, although this is much larger than the bandwidth available for terrestrial signals.

NEW TECHNOLOGY FRONTIERS IN BROADCASTING

Developments in digital coding, modulation, and compression have made the transmission of digital audio commonplace and digital video feasible. These technologies are gradually making their way into the conventional broadcasting scene as digital audio and ATV (Advanced TV). The potential for an enhancement to AM and FM broadcast stations exists where the sound quality can equal or nearly equal that of CD-technology. Also, HDTV in terrestrial mode is being developed in the United States, Europe, and Japan as a means of providing greatly improved picture quality to television viewers. The development of a successful HDTV system will provide the basis for revolutionary new video services to many homes, industry, scientific, and medical organizations. The implementation of HDTV in terrestrial mode is difficult and expensive for broadcasters, but it appears essential for broadcasters to find a way to upgrade their facilities to provide HDTV to consumers and remain competitive with the virtually certain introduction of HDTV by cable, VCR's, and DBS. International broadcast stations will experience improved spectrum efficiency with the planned single-sideband implementation and with satellite-sound broadcasting potentially representing a supplemental delivery system to international audiences. The technology has reached to the point where digital sound and picture are broadcast by maintaining quality and cost. A technology demonstration of all-digital production and transmission was done in the International Broadcasting Convention (IBC) in 1995, held at Amsterdam[2]. The DDS (Digital Delivery System) was capable of providing up to 16 simultaneous stereo replay channels in either linear or MPEG-2 format. Thus, broadcasting in recent times is undergoing a phenomenal change.

Broadcast technology is a complex mix of many technologies. It encompasses audio/video engineering, acoustics and electrical engineering, electronic, air conditioning, satellite, digital and many more upcoming techniques. In this Chapter certain basics of broadcast technology shall be described.

COMMUNICATION SYSTEM

As the technological development of broadcast system is discussed, it would be seen that the basic communication system has remained the same with three distinct components viz. Transmitter, Channel and the Receiver Fig. 2.1. The functions of these components are described below:

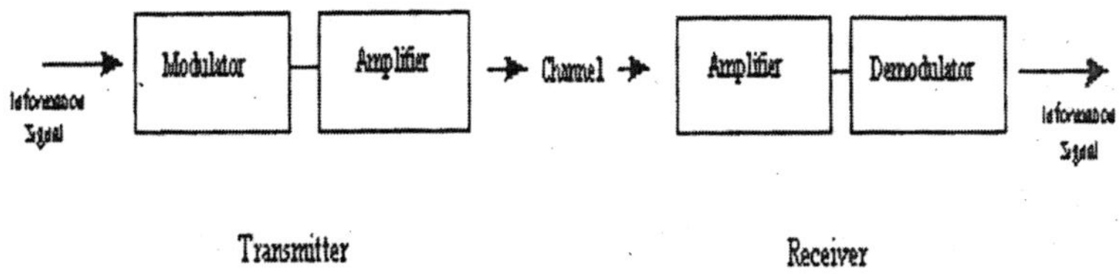

Fig.2..1 Basic Communications System.

Transmitter: The sub-system that takes the information signal and processes it prior to transmission. The transmitter modulates the information signal onto a carrier signal, amplifis the signal and broadcasts it over the channel.

Channel: The medium, which transports the modulated signal to the receiver. Air acts as the channel for MW (Medium Wave), SW (Short Wave) and FM (Frequency Modulated) radio broadcasts. Modern broadcast also uses a number of other mediums like cable TV or the Internet.

Receiver: The sub-system that takes in the transmitted signal from the channel and processes it to retrieve the information. The receiver must be able to discriminate the signal from other signals which may be using the same channel, amplify the signal for processing and remove the carrier to retrieve the information. The desired signal is selected against unwanted signals in the channel, by a method known as tuning. Similarly, the retrieving of the information signal is accomplished by a technique known as demodulation.

HOW TELEVISION WORKS

The heart of television is the method used to convert images to electronic signals, transmit it from one location to the other and recover it back at the receiver. The process is the same irrespective of the delivery method viz. radio waves, coaxial cable or fiber optics.

The first part of the process involves getting the image into an electronic form. The picture is broken up into small pieces by a scanning process, and sent one line at a time, with each line containing hundreds of bits of the picture. This is called the video signal. In early cameras, various forms of special tubes were used for this process. The tubes had a special layer that was sensitive to light. As an electron beam scanned this layer, the image that was focused on the tube was converted to an electronic voltage, which corresponded to the brightness of the image Fig. 2.2. Modern cameras use solid state devices that do essentially the same thing. Colour images are sent in the same way, except that there must be a signal for each of the primary colours used.

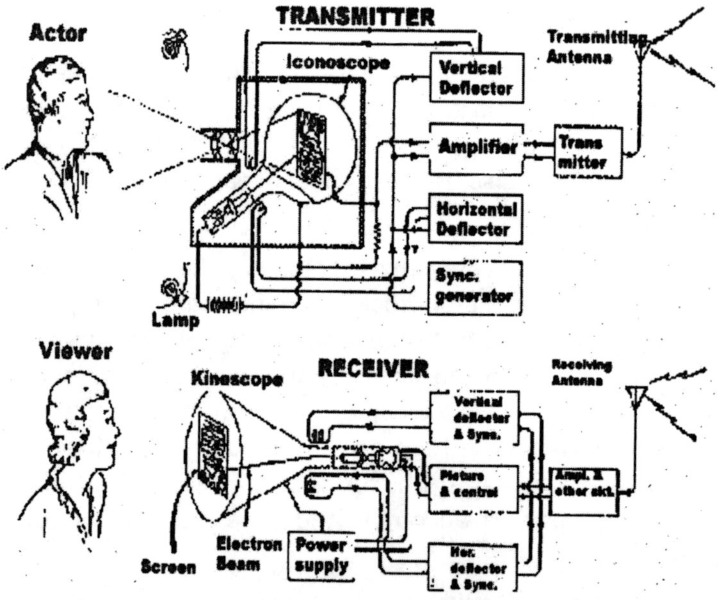

Fig. 2.2. Television transmission and reception system.
Source : RCA television handbook, 1939.

Regular television stations use a form of AM modulation known as vestigial sideband (VSB) to send the video via radio waves. Similar to AM, VSB filters out duplicate parts of the modulated signal in order to reduce the amount of channel space, or *bandwidth* necessary to send the signal. The signal is sent through terrestrial transmission systems using VHF or UHF antenna or satellite system. Traditional satellite TV uses microwave frequencies that require large dishes for best performance Fig. 2.3. These are the systems that have been available to consumers since the late 70s. These systems use frequency modulation (FM) for sending the video, which reduces interference between the various satellite transponders (satellite circuitry which relays the TV signal). The new DSS (Direct Satellite System) satellite TV units operate at a higher frequency than traditional units, which allows the dishes to be smaller. In addition, the video signals are converted to digital data, which is sent over the satellite link. At the receiver the digital signals are converted back to video which is sent on to the TV set. DSS signals are generally of higher quality than conventional satellite TV.

Fig. 2.3. Satellite Dish Antenna.

THE VIDEO SIGNAL

Scanning

The video signal consists of several smaller elements, each having a distinct function. Video is line-by-line information of the image. Television pictures are made up of horizontal lines. Each line is transmitted one at a time in a process called scanning. In India, each picture remains on the screen for 1/25th of a second and is made up of 625 horizontal lines. These lines are created one at a time on a picture tube by an electron beam. When the beam has sprayed 625 lines, the viewer sees one still frame of a video picture. The process is repeated 25 times each second to create illusion of motion. Each of the 25 frames is a still image, but each shows a progressively different stage of the motion. It is really like watching a slide show in fast motion. 25 stills are displayed every second, but they produce the illusion of motion due to "persistence of vision." Without it, neither motion pictures nor video would exist. The scanning electron beam starts at the top left of the picture tube and writes one horizontal line. When it reaches the right hand side of the picture or raster area, the beam drops down and writes the next line from left to right Fig. 2.4. After one field has been scanned, the next field's scanning starts at the top middle of the picture and proceeds horizontally as before. This time however, the picture tube scans between the lines scanned during the previous field.

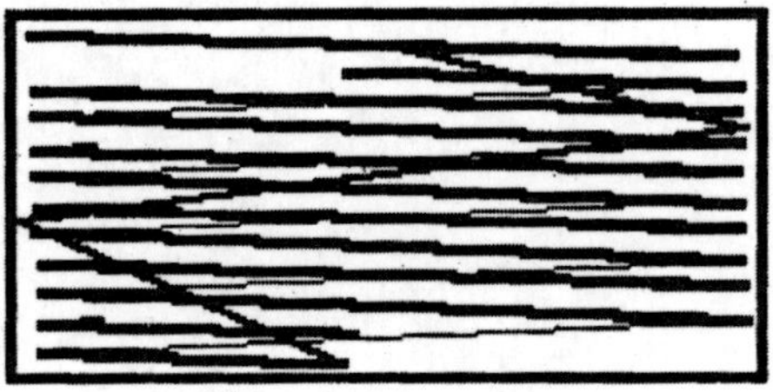

Fig. 2.4. Scanning Process.

Interlaced Scanning

In the early television systems, this process of writing 625 lines for each frame created noticeable flickering. To minimize the flicker, a system of "interlaced scanning" was developed. The interlace system divides each frame into two separate fields each with half of the picture information for a total of 625 lines of picture information. The first field contains all odd-numbered lines #1, #3, and so on. The second field contains the even-numbered fields #2, #4, etc. After field a is scanned for all the odd-numbered lines, a vertical synchronization pulse returns the electron beam to the top centre of the picture tube and then scans all of the even numbered lines. Each of the 25 frames of a video picture includes these two interlaced fields so the actual scanning rate is 50 fields per second. If the scanning were not interlaced, the entire 625 lines would illuminate the screen 25 times per second. The eye starts to blend flashing pictures when the pictures scan at a rate of about forty-five per second. This gives the viewer the impression of a continuously lighted image. If the lines are not interlaced, the resulting 25 per second scan rate would flicker and would be objectionable to the eye.

Blanking

In order so that the retrace lines do not show up and possibly "fog" the picture, a Horizontal Blanking signal is added to the camera video output. This happens at the end of each horizontal scan. To prevent the vertical retrace lines from showing up across the picture as a series of diagonal lines, a Vertical Blanking signal is added at the end of each field Fig.2.5.

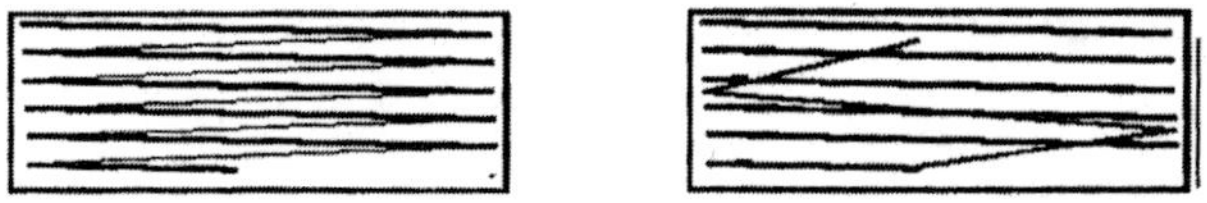

Fig. 2.5. Blanking Signal.

Synchronizing Signal

When a video picture is displayed or transmitted a synchronizing signal is sent along with camera video signal. Its purpose is to control the TV monitor's scanning circuits. This allows to precisely scan what the camera is scanning.

Horizontal Sync : At the end of each line scan, just after the horizontal blanking signal has started, a Horizontal Sync pulse is included during the blanking interval. It is a "negatively going" pulse of short duration that reverses the beam until it arrives in the proper location to start the active scan.

Vertical Sync: At the end of each field, shortly after vertical blanking has started, a Vertical Sync signal is included. It is a closely spaced series of six wide "negatively going" pulses that are sensed by the vertical scanning section in a TV receiver. They trigger the vertical scan circuits to retrace up and start the next field.

VIDEO IMAGE AT RECEIVING END

At the most basic level, the video image is made up of phosphorescent dots—*pixels* or *phosphors*—that are arranged in horizontal lines on the TV screen. *An electron gun* (three electron guns, in most colour TVs) in the rear of the picture tube, or *cathode ray tube (CRT)*, fires an electron beam at these pixels, scanning line by line across the 525 lines of the TV image. The number of scan lines varies in different countries that use different broadcasting systems. In Europe, for example, the image consists of 100 more scan lines than it does in North America. The U.S. standard was set by the NTSC (National Television Standards Committee) back in 1941. When struck by the beam the pixels glow and thereby create the television image.

COLOUR TV

Red, blue and green are TV's primary colours and yellow, magenta and cyan are considered secondary colours. By combining the proper mixture of red, blue and green light any colour of the rainbow can be produced. Therefore, in colour television only three colours (red, blue and green) are needed to produce the full range of colours in a colour TV picture Fig. 2.6, 2.7, 2.8. The colour TV process is based on the process of separating (in the camera) and then combining (in a TV set) red, blue and green. Since each colour can be broken down into some percentage of red, blue or green, every colour can be expressed in terms of some relative proportion of each of these primary colors. Since white is the presence of all colours, the camera's CCDs responding to pure white at the simultaneous presence of all three colours. Black is simply the absence of all three colours. Unfortunately, the human eye does not see all colours with equal brightness. The eye is much more sensitive to yellowish-green light than to either blue or red light. Due to the greater sensitivity of the eye to the green-to-orange portion of the colour spectrum, an equal, percentage of red, green and blue mix of coloured light does not appear white. Because of this, and also the nature (and limitations) of the colour phosphors used in TV sets, the actual colour mix used in colour television ends up being about 30 percent red, 11 percent blue and 59 percent green.

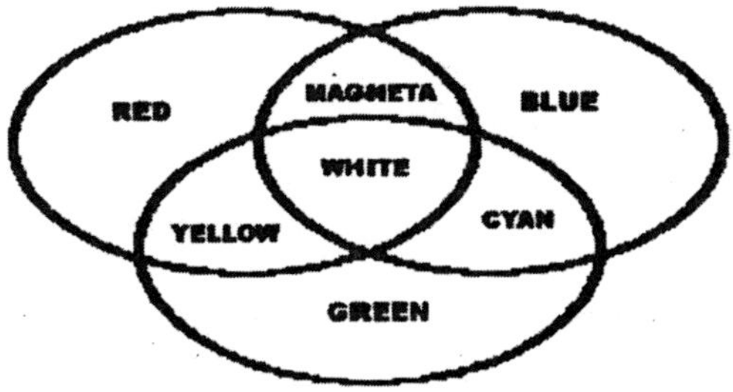

Fig. 2.6. Colour Production in TV.

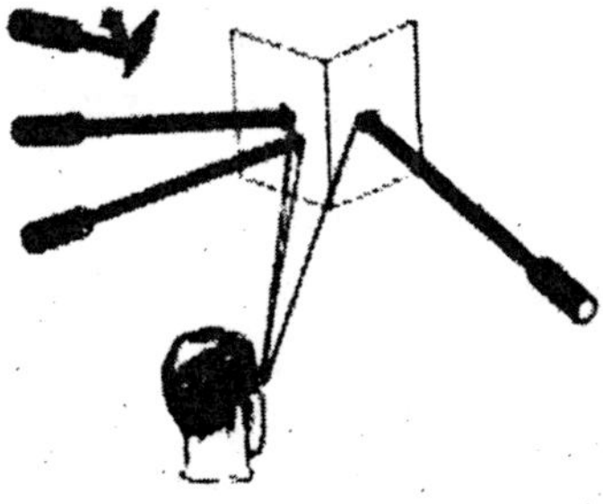

Fig. 2.7. Red Plus Blue.

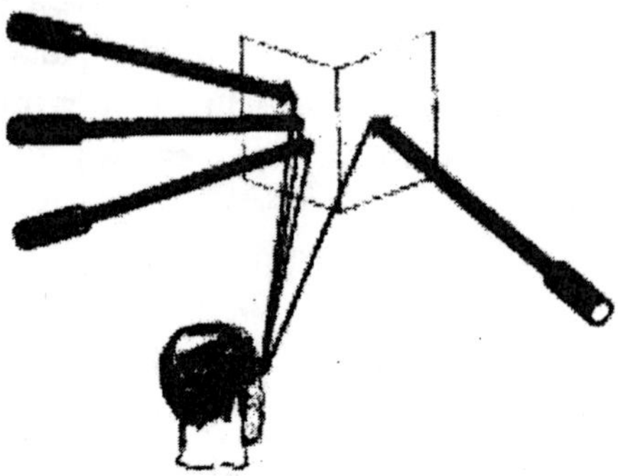

Fig. 2.8. Green Plus Red Plus Blue.

In the camera, the colour is decomposed into primaries by means of prisms Fig. 2.9. Each primary illuminates a separate tube and therefore produces its own signal. In receivers, the colour is reconstituted using a large number of luminescent spots arranged in red-green-blue triplets Fig. 2.10. The spots are so close enough that, from a reasonable viewing distance, a triplet appears as a single information source. In other words, the eye sees each triplet as one picture element Fig. 2.11. The number of discernible colours with television primaries, is around ten thousand.

Video and film making involves different approaches to creating colour images. Video constructs colours by adding them together (additive colour). A single phosphor on the TV screen is coloured red, green, or blue. The electron gun (or guns) ignite three nearby phosphors and combine their individual colours, thus generating a broad variety of colours. Film, in contrast, is a subtractive colour process. As white light from a projector lamp passes through a piece of motion picture film, yellow, magenta (reddish), and cyan (bluish) colours are filtered out of the light. The colours that are not filtered out form the many colours of the spectrum.

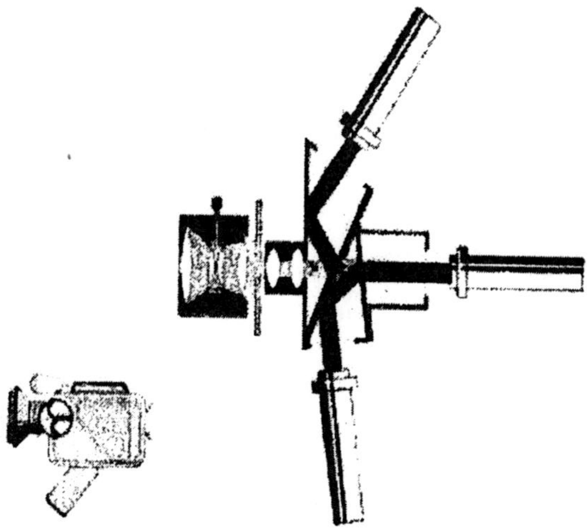

Fig. 2.9. Colour in Camera.

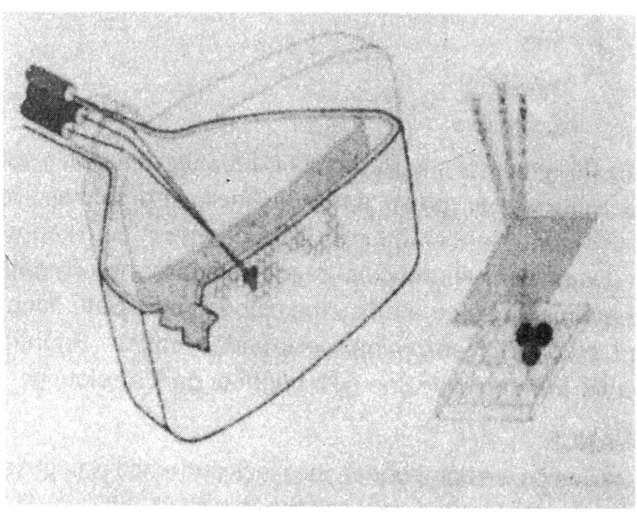

Fig. 2.10. Colour in Receiver.

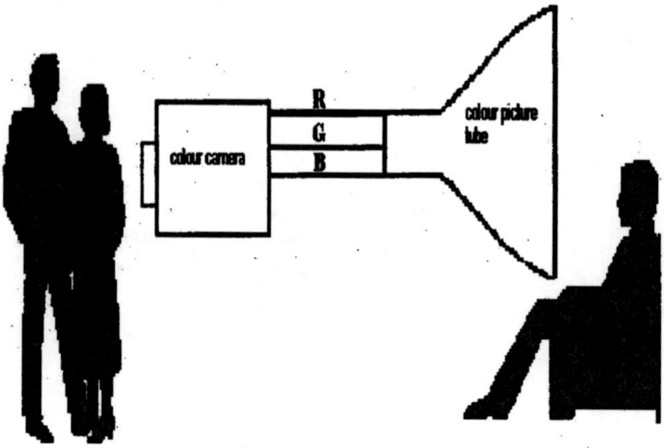

Fig. 2.11. Colour Picture System

IMAGE QUALITY: COLOUR AND BLACK AND WHITE

There are a few basic colour characteristics that are described the same in both video and film:

- hue
- saturation
- brightness.

Hue designates a specific colour from within the visible spectrum of white light, e.g., red, green, and blue. The level of saturation defines a colour's purity—how much or little grayness is mixed with the colour. Deep, rich, vibrant colours are said to be heavily saturated. They become less saturated as the colours fade. Saturation is also termed *chroma* or *chrominance* in video colour. Brightness or *luminance* in video indicates how bright or dark a colour is.

LUMINANCE

Luminance is a component, the black and white or bright element of an image. It is written as Y, so the Y in YUV, YIQ, (Y, R-Y, B-Y) and Y, Cr, Cb is the luminance information of the signal. In a colour TV system, the luminance signal is usually derived from the RGB signals, originating from a camera or telecine, by a matrix or summation of approximately: Y = 0.3R + 0.6G + 0.1B Fig. 2.12.

CHROMINANCE

The colour part of a signal, relating to the hue and saturation but not to the brightness or luminance of the signal is chrominance. Thus, black, gray and white have no chrominance, but any coloured signal has both chrominance and luminance. Cr and Cb, (R-Y) and (B-Y) represent the chrominance information of a signal.

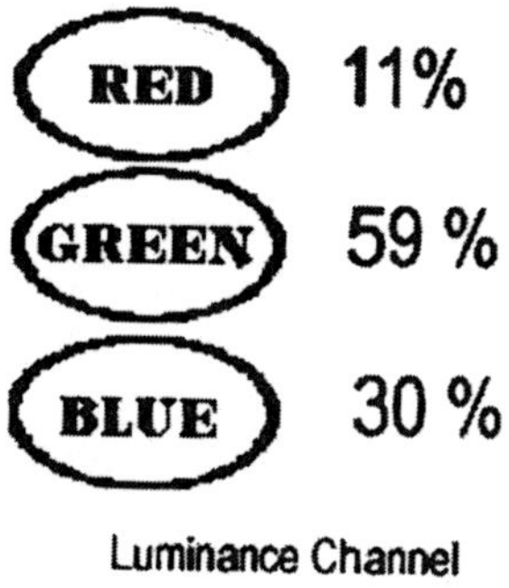

Fig. 2.12.. Colour for Luminance Channel.

COMPOSITE VS. COMPONENT

Component video means colour video represented by three separate signals (luminance and two colour difference signals). This is commonly referred to as Y/Cr/Cb colour space. RGB is almost never used in recording, because it requires 3/2 times the bandwidth of Y/Cr/Cb representation in order to achieve similar subjective quality.

NTSC SIGNAL

In countries that adhere to the NTSC conventions (National Television Standards Committee), there are 525 lines and 60 fields per second. NTSC system uses seven types of electronic information that comprise the video image. Together they are called the television composite waveform, more commonly referred to as composite video.

The Composite Video Signal

The seven elements of the composite video signal include:

- horizontal line sync pulse
- colour reference burst
- reference black level
- picture luminance information
- colour saturation information
- colour hue information
- vertical sync pulseu.

Horizontal Line Sync Pulse

Before each line is scanned, horizontal sync pulses set the electron beam to a locked position so that each line of picture information starts at the same position during scanning.

Colour Reference Burst

To insure standard hue and colour saturation, a 3.58 MHz colour reference burst is added before the picture information on each scan line. It is a sine wave with nine cycles.

Reference Black Level

Black level is also called "setup" or "pedestal." It is defined as 7.5 IEEE units (from the Institute of Electrical and Electronics Engineers). Formerly these units were called IRE units from the Institute of Radio Engineers.

Picture Luminance Information

Picture luminance ranges from 7.5 IEEE units for black to 100 IEEE units for peak white.

Colour Saturation

Colour saturation is overlaid on the picture luminance information. This overlay is called the 3.58 MHz subcarrier. It is compared to the standard colour reference burst. This comparison determines the saturation of colours displayed on the set.

Colour Hue

Colour hue is also present in the 3.58 MHz subcarrier. The accuracy of colours in the picture are determined by the phase or rotation of the colour hue information.

Vertical Sync Pulse

The vertical sync pulse controls the length of time of the vertical blanking interval. This is the period when the TV screen goes blank between the end of one field and the beginning of the second field. The vertical blanking interval is sometimes used for inserting time code, automatic colour tuning and captioning information in the video signal.

There is also a horizontal blanking interval, which occurs between the end of one scan line and the beginning of the next. The horizontal sync pulse controls this blanking interval. Within this interval are the horizontal sync pulse and the colour reference burst.

BANDWIDTH AND LINES

Both of these figures refer to the horizontal resolution. "Lines" refers to the number of vertical alternating black and white lines.

$$BW = \frac{\text{Aspect ratio} * \text{resolution in lines}}{\text{Length of active picture line}} / 2.$$

ASPECT RATIOS / SAFE TITLE AREA

The aspect ratio is the width-height proportion of the picture. All the TV standards use the same 4:3 aspect ratio irrespective of their number of scanning lines. The 4:3 ratio was consistent with motion pictures that predated the wide screen aspect ratios used in Cinemascope, Vista-Vision and Panavision films. HDTV/DTV's use 16:9 aspect ratio Fig.2.13 which more closely conforms to these widescreen processes. Due to image alignment differences on home receivers, 15 per cent around the edge of the image is considered unsafe to place any important information. The remaining 85 per cent of the image is considered safe title area.

Fig. 2.13. Safe Area.

The differences between film and television aspect ratios are most apparent in anamorphic films, but they are also evident in the transfer of masked widescreen films to videotape. When masked widescreen films—with an aspect ratio of 1.85—make the transition to TV's aspect ratio of 1.33, they lose a little from the edges, but not much because of the technique used to create this form of widescreen. The masked widescreen films use the entire 1.33 frame when shooting, but blacken the tops and bottoms when projected in a theater.

TELEVISION STANDARDS

Twenty years ago, few hundred languages and a dozen or so incompatible systems of television in the world did not matter much since distance was a great insulator. However, with the development of satellite and the Internet, incompatible standards and incomprehensible languages are not acceptable since they create barriers to communication and understanding. Broadcast television standards refer to technical approaches to broadcasting the picture and sound. These are required to be uniform so that a programme produced in one country can be automatically viewed in another country without any need for it's conversion to the appropriate technical standard. There have been about 14 different broadcast standards in use at different times throughout the world. Over a period of time a lot has been accomplished to facilitate easy reception of foreign broadcasts by simplifying television standards. The discontinuation of the high-definition 819 lines French standard and the rather poor 405 lines British standard left the world with just two standards; 525 lines (generally accompanied by the American NTSC colour system) and 625 lines (usually in PAL or SECAM colour). Today, excluding HDTV/DTV (high-definition TV/digital TV), three basic systems serve the vast majority of countries. The difference between these basic international broadcast standards are primarily based on:

- The number of horizontal lines in the picture
- The broadcast channel bandwidth of the signal
- Type of modulation viz. AM or FM, used for the audio and video.

NTSC

The United States National Television Systems Committee, a broadcast engineering advisory group established the NTSC (television standard). The broadcast standard for colour television system in the USA, Canada, Mexico and Japan is NTSC M (M defines the 525/60 line and field format). Generally, the standard is just referred to as NTSC. The bandwidth of the NTSC system is 4.2 MHz for the luminance signal and 1.3 and 0.4 MHz respectively for the R-Y and B-Y colour difference signals modulating the I and Q axes of the NTSC subcarrier.

NTSC/525 system has the advantages of higher frame rate that reduces flicker, possibility of edit at any four field boundary point without disturbing the colour signal and less inherent picture noise. The main disadvantages are—reduced clarity on large screen TVs due to lower number of scan lines, smaller luminance signal bandwidth, susceptibility to hue fluctuation, etc.

PAL

PAL i.e. Phase Alternating Line is the colour coding system for television widely used in Europe and throughout the world, almost always with the 625/50 line per field system. It was derived from the NTSC system but by reversing the phase of the reference colour burst on alternate lines (Phase Alternating Line). PAL is able to correct hue shifts caused by phase errors in the transmission path. The bandwidth of the 625/50 PAL system is 5 or 5.5 MHz for the luminance signal and 1.3 MHz for the R-Y and B-Y colour difference signals modulating the V and U axes of the PAL subcarrier.

PAL/625 has the advantage of greater number of scan lines resulting in more picture detail. Further placing of the colour subcarrier at 4.43 MHz results in wider bandwidth of luminance signal that allows a larger bandwidth of monochrome information to be reproduced than with NTSC/525. The other advantages are stable hues due to reversal of subcarrier phase on alternate lines, any phase error will be corrected by an equal and opposite error on the next line, correcting the original error. Moreover the gamma value for PAL/625 is set at 2.8 as opposed to the lower 2.2 figure of NTSC/525. This permits a higher level of contrast than on NTSC/525 signals. This is particularly noticeable when using equipment with multi-standards, as the contrast and brightness settings need to be changed to give a similar look to signals of the two formats.

PAL/625 suffers from noticeable flicker due to the lower frame rate, lower signal to noise ratio, loss of colour editing accuracy due to the alternation of the phase of the colour signal, variable colour saturation, etc.

SECAM

The french designed a system of their own, primarily for political reasons to protect their domestic manufacturing companies which is known as SECAM, standing for Sequential Couleur Avec Memoire. SECAM was widely adopted in Eastern bloc countries to encourage incompatibility with Western transmissions. SECAM shares with PAL the ability to render images with the correct hue, and goes a step further in ensuring consistent saturation of colour as well. SECAM also shares with PAL/625, the higher number of scan lines than NTSC/525.

SECAM suffers from greater flicker like PAL/625. Mixing of two synchronous SECAM colour signals is not possible and as such most TV studios in SECAM countries originate in PAL and transcode prior to broadcasting. More advanced home systems such as Super VHS, Hi-8, and LaserDisc work internally in PAL and transcode on replay in SECAM market models. The other problems are patterning effects even on non-coloured objects, lower monochrome bandwidth incompatibility between different versions of SECAM.

D-MAC

In Europe, a few Direct Satellite Broadcasting services use a system called D-MAC. It's use is not widespread at present and it is transcoded to PAL or SECAM to permit video recording of it's signals. It includes features for 16:9 (widescreen) aspect ratio transmissions and an eventual migration path to Europe's proposed HDTV standard. There are other MAC based standards in use around the world including B-MAC in Australia and B-MAC60 on some private networks in the USA. There is also a second European variant called D2-MAC which, supports additional audio channels making transmitted signals incompatible, but not baseband signals.

PALPLUS STANDARD

The PALplus standard generates for each TV frame, 432 (216 per field) main active lines in 16:9 format, plus 144 (72 per field) helper active lines. The main active line contains a conventional PAL image information, while the helper lines contain additional information transmitted in a way that a conventional PAL TV receiver will interpret them as "black" lines. Those lines are transmitted for each field in the sequence: (i) a group of 36 consecutive helper lines, (ii) 216 main lines and (iii) the remaining 36 helper lines. When receiving such a signal, a conventional 4:3 PAL TV receiver will display the image in 16:9 format in the centre of the screen (letterbox format, already widely used for broadcasting feature films). A PALplus TV set generates a full-screen image by appropriately combining main lines with the helper lines.

The main video signal standards and the technical characteristics of the various transmission systems are given in Table 2.1. The colour standards (PAL, SECAM, NTSC) and television system used by various countries are indicated in Appendix 1 of this book.

Table 2.1. Technical Characteristics of the Various Transmission Systems

Name	*Frame/Field Rate*	*Aspect Ratio*	*Scan Lines*	*TV standard*	*Colour System*	*Subcarrier Frequency*
NTSC	29.97/59.94	4:3	525	EIA	NTSC	3.58 MHz
PAL M	29.97/59.94	4:3	525	EIA	PAL	3.58 MHz
SECAM-M	29.97/59.94	4:3	525	EIA	SECAM	4.43 MHz
PAL	25/50	4:3	625	CCIR	PAL	4.43 MHz
SECAM	25/50	4:3	625	CCIR	SECAM	4.25/4.40 MHz
D-MAC	25/50	4:3 or 16:9	625	D-MAC	D-MAC	N/A
PALplus	25/50	16:9	625	CCIR	PAL	4.43 MHz
HiVision	60/120	16:9	1125	HiVision	MUSE	Unknown

ADDITIONAL FEATURES

In addition to standard combinations of scan rate, colour system and transmission frequencies, there are additional features like Stereo sound, subtitling and information services that differ from system to system. In the cases of both stereo sound and additional textural information carried in the top few lines of the picture, there are three competing systems of varying technical merit. These are described below:

MTS

The oldest, still operational of the stereo sound systems is the American MTS system based on NTSC transmissions. It is used in conjunction with NTSC/525. It consists of two independent carriers each carrying a discrete channel. One channel provides stereo sound by providing left/right channel difference signals relative to transmitted mono audio track. The second carrier carries the secondary audio program (SAP) which is used for a second language or a descriptive commentary for the blind. MTS uses a technique based on the dbx noise reduction to improve the frequency response of the audio channel.

FM-FM

A recent development is the twin channel FM-FM system used in Germany, Austria, Australia, the Netherlands and Switzerland. FM-FM is analog system with dual carrier FM coded discrete stereo transmissions. It has the capability for being used for bi-lingual operation under user selection, but no auto-selection is available. Audio quality is better than standard mono soundtrack.

NICAM

BBC designed the most recent system, NICAM 728, in the late 1980s using digital audio technology. NICAM 728 commonly called NICAM stands for Near Instantaneously Companded Audio Multiplex. It uses digital two-channel audio transmissions with sub-code selection of bi-lingual operation. Stereo digital signals with specifications approaching those of Compact Disc are also possible. NICAM uses a 14 bit sample at a 32 KHz sampling rate, which produces a data stream of 728 KBits/sec.

ADDITIONAL INFORMATION

TV signal also carries auxiliary information. Few of these are mentioned below:

Subtitling

CC - (Closed Captioning)

The U.S. Closed Captioning mechanism came about through political pressure from the Deaf Organizations in the USA and has not been developed beyond the simple job of producing subtitles for the Deaf. CC contains subtitling information only and is transmitted on line 21 of NTSC/525 transmissions. CC has no support for block graphics or multiple pages but it can support 8-colours and the use of an italic typeface. Frequently found on pre-recorded VHS cassettes and LDs, also used on broadcast. CC is also used on PAL/625 pre-recorded VHS cassettes in a modified version.

TeleText

The oldest of the subtitling systems is almost definitely the BBC and IBA designed TeleText system which has been is use in the UK since mid-1970s. It is also the most widespread and the most flexible of the systems in widespread use. TeleText is an information service of 200-700 "pages" covering a wide range of topics including TV Schedules, News, Financial Market Prices, Comment, Reviews, Concert and Theater information. Subtitles are typically transmitted on page 888 in the UK, on pages 199/299/399 in Belgium and Holland, on page 150 in Germany and on page 777 in Italy. There are a number of variant character sets used, but the encoding is identical. Includes support for 8 colours, and limited block graphics, and selective revealing of underlying TV picture. Transmitted on a variable number of lines (specified in header which contains basic information such as time, date and channel), starting on line 12 and continuing for 7-8 lines typically. Found on broadcasts and some laser discs; recording of TeleText signals is marginal on S-VHS, almost impossible on VHS hence the PAL/625 version of CC.

The BBC-designed TeleText system is known in some quarters as World Standard Tele Text. There is an enhanced version known as Fastext, which defines four links to additional pages that can be followed with one of four coloured buttons on the Tele Text receiver's remote control. In the U.S., CC is known widely in technical circles as LITO (Line Twenty-One).

ANTIOPE

The French developed a subtitling and information system called Antiope which has not found favour elsewhere, largely due to the existing widespread use of the BBC developed TeleText system.

The technical specifications for various transmission codes are given in Table 2.2.

Table 2.2. Technical Specifications of Transmission Standards.

Code	*Frames*	*Scan Lines*	*Freq. Band*	*Sound Offset*	*Vision Modulation*	*In Use*
Terrestrial Transmission Standards						
A	25	405	VHF	-3.5MHz	Pos	No
B	25	625	VHF	+5.5MHz	Neg	Yes
C	25	625	VHF	+5.5MHz	Pos	Yes
D	25	625	VHF	+6.5MHz	Neg	Yes
E	25	819	VHF	+11MHz	Neg	No
F	25	819	VHF	+5.5MHz	Pos	No
G	25	625	UHF	+5.5MHz	Neg	Yes
H	25	625	UHF	+5.5MHz	Neg	Yes
I	25	625	UHF	+6.0MHz	Neg	Yes
K	25	625	UHF	+6.5MHz	Neg	Yes
Kl	25	625	UHF	+6.5MHz	Neg	Yes
L	25	625	UHF	+6.5MHz	Pos	Yes
M	30	525	VHF/UHF	+4.5MHz	Neg	Yes
N	25	625	VHF/UHF	+4.5MHz	Neg	Yes
Satellite Transmission Standards						
Ku-Band	Any	Any	~11GHz	+6.50MHz	Neg	Yes
C-Band	Any	Any	~4GHz	+6.50MHz	Pos	Yes

STANDARDS CONVERSION

Standards conversion is the most complex and degrading of all of the techniques for building bridges between TV standards.

Conversion between TV systems with same frame rate is usually reasonably effective, except of course that the resultant signal usually suffers from the deficiencies of both the source and destination colour system. Conversion between TV systems with different frame rates is extremely difficult and there is no perfect solution. With a trained eye, it is possible to spot that any given piece of material has been standards converted within a matter of seconds. The units that perform these conversions are known as Digital Standards Converters.

BASICS OF DIGITAL TV

Digital television technology is an extension of computer and image processing technology. Advantages are easy storage and great scope for image processing. The values of the brightness or colour of picture elements along a television line can be represented by a series of numbers. If these are expressed in base 2, each value can be transformed into a sequence of electrical pulses.

HOW DIGITAL SAMPLING WORKS

The operation which converts from the "analogue" world to the "digital" world comprises of two stages:

- sampling in which the value is measured at regular intervals and
- quantification in which each measurement is converted into a binary number.

These operations are carried out by an analogue to digital converter. The series of "1" and "0"s obtained after quantification can be modified (i.e. coded) to counteract more effectively the disturbances the signal may meet during transmission.

Since the signal has only two possible values (0 or 1), detection is based on the presence or absence of the signal. Hence the possibility of regenerating it.

ADVANTAGE OF DIGITAL SYSTEM

The signal can be preserved during successive recordings or on noisy transmission paths. This technique is already in widespread use for special effects on existing images. It lies at the root of computerized image synthesis systems.

EVOLUTION OF DIGITAL BROADCAST SIGNAL FORMATS

PAL, SECAM, and NTSC were adopted for composite colour signals in 1960s for maintaining a compatible reception by millions of black and white TV sets. To keep the same channel allocation and to save the frequency spectrum, the same black and white channel bandwidth was used; approximately 4.2 MHz for NTSC and 5.5 MHz for PAL/SECAM. Such a low bandwidth was found adequate for transmission to the TV receivers, albeit with drawbacks like "cross colour" in specific conditions. However, these were not sufficient for processing in the studios; i.e. for chroma keys. This led to the evaluation of component standards in 1981, which are commonly known as "Betacam", and the "M" formats. This analog component standard had a disadvantage in the post-production—that it needed three cables for every signal instead of one for the composite. For all these reasons, it was necessary to use a digital television standard that would make it possible to preserve the original quality whatever the processing complexity. CCIR 601 (usually referred to as 4:2:2) was defined as an international standard for component coding of TV signals. It specified orthogonal sampling at 13.5 MHz for luminance (Y) and 6.75 MHz for the two-colour difference signal Cb and Cr.

Some extended definition TV (EDTV) systems use a higher resolution format called 8:4:4 which has twice the bandwidth of 4:2:2. The choice of sampling frequencies gives 720 samples / active line for luminance Y and 360 samples for colour differences and also includes space for representing analogue blanking within the active line. A serial digital signal for carrying the signal on a single coaxial line was developed in 1984. The system was standardized for 243 Mbits/sec based on 8/9 bit transcoding. Subsequently a serial digital component standard based on 10 bit was developed.

ITU-R 601

This standard defines the encoding parameters of digital television for studios. It is the international standard for digitizing component of television video in both 525 and 625 line systems. It is derived from the SMPTE RP125 and the EBU Tech. 3246-E. ITU-R 601 deals with both colour difference (Y, R-Y, B-Y) and RGB video, and defines sampling systems, RGB/Y, R-Y, B-Y matrix values and filter characteristics. It does not actually define the electro-mechanical interface. ITU-R 601 is normally taken to refer to colour difference component digital video (rather than RGB), for which it defines 4:2:2 sampling at 13.5 MHz with 720 luminance samples per active line and 8 or 10 bit digitizing. Some headroom is allowed with black at level 16 and white at level 235—to minimize clipping of noise and overshoots. Using 8 bit digitizing approximately 16 million unique colours are possible. The sampling frequency of 13.5 MHz was chosen to provide a politically acceptable common sampling standard between 525/60 and 625/50 systems, being a multiple of 2.25 MHz, the lowest common frequency to provide a static sampling pattern for both.

8 BITS OR 10 BITS

There has been considerable debate about the bit resolution that should be used for digital television and film images. The CCIR 601 digital television sampling standard specifies that both luminance and chrominance information in a picture can be sampled at 8 or 10 bits. Originally it specified 8 bits—a recommendation based on many years' of study of television signals and parameters. Analogue pictures were digitized and converted back to analogue via such machines as synchronizers, standards converters or digital effects systems and the 8 bit system appeared to work well. It was only after some experience in processing the signal that some contouring artifacts were noticed that was objectionable. The use of 10 bits was thought to be a possible solution. 8 bits digitization allows maximum 256-(28) levels of information to be represented for each component (Y, Cr. and Cb). The system architecture and signal processing need attention and careful design. The first criterion has to be that the digital system must not degrade the image in any noticeable way. This means the quality of the digital system must be at least a match for the rest of the signal chain. The latest CCD cameras are quoted as offering 60 dB S/N (signal to noise ratio), a figure that can be met by 8 bit digital systems. Looking at the other end of the chain, the viewer typically receives at home between 25 dB to 45 dB S/N. Clearly there is still room for some improvement in the transmission/reception chain. Digital graphics systems and computers are able to produce pictures that are not technically the same as those produced by a camera. The 10 bit solution attempts to correct the symptoms, but it does not address the cause: The logical conclusion in this approach would require not 10, 12 or 14 bits, but 16 bits to eliminate the problem. This is totally impractical. A solution—known as Dynamic Rounding was developed initially for applications with print products. This process can truncate the 16 bit words to 8, taking account of the bits below the top 8, to weigh the amount of least significant bit (LSB) dithers. This effectively allows levels between the LSBs to be displayed. The result is perfectly smooth; there is no contouring. Dynamic Rounding not only effectively negates the need to go to 10 bits, but it can actually produce better results. The artifacts can appear when the word length is truncated, and as such attention must be paid when including a mix of 8 and 10 bit equipment in the digital production and post-production chain. Going from 8 bit to 10 bit is no problem but from 10 to 8 will necessarily involve truncation. Inputs should be equipped with the necessary rounding process otherwise artifacts will be produced. The Dynamic Rounding process is the key to solving this type of contouring problem where word lengths of digital images have to be truncated.

COMPRESSION (VIDEO)

Compression is the process of reducing the amount of bandwidth or data rate for video. The broadcast standards used today—PAL, NTSC and SECAM—are analogue video compression systems. For digital systems pictures are analyzed looking for redundancy and repetition and unnecessary data is discarded. The techniques were primarily developed for digital transmission but have been adopted as a means of handling digital video in computers and reducing the storage demands for digital VTRs.

COMPRESSION, THE WAY OF THE FUTURE?

Digital compression techniques are used to reduce the very large storage requirements and data speeds required for CCIR 601 digital video. Their applications include allowing PCs to handle video (as seen with off-line editing), digital transmission and in some digital VTRs as a means of both reducing the required data rate and increasing capacity. The quality of the results varies widely. For digital equipment, using CCIR 601 has become a mark of the highest quality, to the point where quality issues disappear, but they are re-opened by compression.

Compression is not new to TV. The PAL, NTSC and SECAM colour systems were all devised to compress the full bandwidth R, G and B signals from cameras and telecines into a single 5.5 or 4.2 MHz channel. Experiences with such signal shows they are generally good for recording and transmission but have a more limited use in production and especially post, production. Likewise-digitally compressed pictures suffer some loss or degradation to the extent that many have been ruled out of their use in post-production.

Digital image compression is relatively new. Digital compression analyzes pictures to resolve them into a series of patterns and colours. In some applications, the quality of the pictures is not too much of an issue. For example, in off-line editing many regard it as only necessary to recognize the pictures to be able to make cutting decisions. In others, especially those which, are in the main picture path from origination to final viewer, quality is carefully supervised. So how can the quality of compressed pictures be measured? At the moment there is no instrumental test. It is up to the eye to look at the pictures and make up the mind. But the other problem is that different pictures produce different quality results. It may happen that a system may produce quite acceptable results on slightly soft and clean pictures, sharp and noisy footage may look far worse. Further, first generation compression may be acceptable but second, third, and fourth generations may be totally unacceptable. If the same pictures are subjected to more than one type of compression the result may be unpredictable. Some have claimed 'Betacam quality' but it is clear from the above that such a statement cannot be taken literally. MOSAIC (Methods for Optimisation and Subjective Assessment in Image Communications) is a European Union project with one of its aims being to produce test picture sequences and define assessment methods for subjectively analysing compressed and processed digital video systems.

No techniques have been developed for directly processing compressed pictures. With intra frame compression, using only the data from one frame, as in JPEG, cuts can be made directly but a dissolve, DVE or any other effect requires the picture to be decompressed and the result re-compressed. If the compression is inter frame, taking data from several frames, such as MPEG, then even cuts will require the data to be decoded and re-coded. Not only does this raise the issue of multi-generation quality just as it did with analogue VTRs, but it means the appropriate coder and decoder must be available.

It is clear that using compression adds significant equipment overheads so there have to be strong reasons for using it. In early applications, compression provided a way for PCs to handle video. They are incapable of handling the continuous data rate of 21 Mbytes/second needed for CCIR-601 video—the world's digital television standard. The cost of non-compressed disk storage makes them an unattractive proposition for off-line use in nonlinear editing. In such applications, intra field compression ratios typically of between 10:1 and 100:1 are used. In some cases only a few TV fields are displayed per second.

Both Digital Betacam and DCT VTRs use compression to reduce the quantity of video data to be recorded onto tape. Here the reduction is quite mild, at around 2:1, and excellent results are obtained but neither VTR system is instrumentally perfect. The user has a choice from absolute quality of uncompressed recorders, D1 or D5 on tape, or disk based systems to various levels of compression to fit applications. For example, news footage is often shot on small format cameras and in conditions not best suited to high quality results. News box uses a small amount of compression to provide on-line nonlinear editing of news where speed and flexibility are especially important. The one area that is destined to rely on compression is that of TV transmission either by cable, fibre, or over the air by satellite or terrestrial means. Here there are real advantages, in that several digital compressed channels can be transmitted in the bandwidth of one analogue channel. For HDTV it is compression that will be used.

TELEVISION LIGHTING

There are six objectives for television lighting:

- To fulfil the technical requirements of the TV system
- To provide 3-D perspective
- To direct attention to important elements in the scene
- To establish the mood of the scene
- To fix the time of the action
- To contribute to the overall aesthetic composition of a shot and sequence of a scene.

COLOUR TEMPERATURE

Light sources produce a mixture of hues. Although different light sources may look "white", some are redder or bluer than others. The fluorescent light is bluer or higher in colour temperature than the tungsten lamp. Colour temperature is the relative amount of reddish or bluish colour quality of light. As the light source becomes bluer in colour, its colour temperature increases and as it becomes redder its temperature decreases. The colour temperature affects the colour of the subject lit by the light source. Colour temperature measures the degree of colour and is expressed in Kelvin degree.

In television production, there are two basic lighting conditions: (1) studio/indoor lighting and (2) daylight. Lighting used in TV is calibrated to one of these two situations. Indoor/studio lighting is calibrated to 3200°K. Outdoor lights are calibrated to 5600°K. The camera must be adjusted to operate in different lighting conditions. Most pro-cameras have a filter wheel located behind the lens. The wheel must be set to either 3200°K or 5600°K depending upon the place where the camera is operated.

Another switch is the white balance. This switch changes the CCD or tube's sensitivity to colours. Every time the light conditions change (i.e. moving the camera indoors from outside) white balance has to be calibrated. To calibrate a white piece of paper is placed in front of the camera while holding down the white balance button. Most new cameras display a message in the viewfinder indicating that the white balance is being adjusted and will also indicate that the camera has finished adjusting.

LIGHTING INSTRUMENTS

A number of lighting instruments are used in TV production.

Spotlights

Fresnel Spotlight

Also known as the "Frezzie", this spotlight is most common in TV studios because of the quality of its output and its ability to be adjusted in any situation. These are available for 750, 1000, and 2000 watts.

Ellipsoidal Spotlight

Most studios have one or a few of these spotlights. These are used for projecting patterns against a background. They are ideal for filling up negative space in a plain background. The patterns are interchangeable.

Lensless Spotlight

Lensless spots, as the name suggests, unlike the Frezzie do not use lenses to control the intensity of light being shined. They are much lighter and more efficient than lens spots and cheaper to operate. They are best suited for remote production.

HMI (Halogen-Metallic Iodide)

HMI is designed for outdoor use. HMIs use halogen bulbs compared to quartz bulbs used by the Frezzie. The HMI lamp is a popular choice for exterior and interior lighting where a match to daylight is needed (about 5600°K). It is more efficient than colour-corrected tungsten (normally about 3000°K) and the lamps run a lot cooler for the equivalent output power. On the downside, they are not dimmable, they normally require a separate ballast, they can generate acoustic noise and the colour of the output varies with age which can be a problem when matching several in one scene. However on balance, they are preferable to tungsten for non-studio lighting and where lighting changes are not required.

PAR (Parabolic Aluminized Reflector)

The PAR unit consists of various lamps, lenses, and reflectors. They are very bright and used in large indoor areas such as a stadium. They can be balanced for indoor or outdoor light settings.

Floodlights

Scoops

The scoop is the most common fill light. It reduces the overall contrast and raises the overall light level of the scene. They range from 1000 to 1500 watts.

Soft Light

The soft light has a square or rectangular opening with a curved reflector surface. They are great for softening eye shadows and smoothing wrinkles. Range from 750 to 5000 watts.

Cyc Light

The cyclorama or cyc light is used to light the cyclorama found in most TV studios. They are used to provide even illumination against the background.

Three Point Lighting

Three-point lighting refers to the use of the three essential lights needed for lighting subjects: the key, fill, and back light.

Key Light

The key light is the principal source of illumination. The fresnel spot is best designed for key lighting. It is used as the base reference for setting the intensity and positions of the other lighting instruments.

The key light is hung from the grid on the studio ceiling. It is placed at 30° to 45° vertical angle next to the side of the camera. As the key light is brought closer to the subject, it washes out the facial textures (i.e. wrinkles). As the key lights horizontal and vertical angles change, so does the skin texture of the talent.

Fill Light

The fill light is used to fill shadows caused by the key light. The intensity of the fill light is about the same as the key's intensity. As the key-to-fill ratio is increased, shadows become prominent. Such high ratios are ideal for dramatic productions. News productions tend to keep the key-to-fill ratio about the same.

Back Light

The goal of the back light is to provide a 3-D perspective in the 2-D TV image. The back light is used to produce a rim of light around the subjects head. Heavy back lighting is popular for nighttime scenes, but too much lighting can create artificial effects. The main instrument for creating the back light is the fresnel spot.

LIGHTING SITUATIONS

Limbo and Cameo Lighting

Limbo lighting, lights talent against a neutral, undefined background. A popular variation, known as cameo lighting, lights talent against a completely black background. A black velour curtain is used to absorb the light to create the effect. Only spotlights can be used to create the cameo effect; any "loose" light that shines on the curtain will ruin the effect.

Lighting with Color

"Gels" are popular colour media used to add punch to a set. Gels, used in TV production are made of either acetate or polyester. Some gels can reduce the light output by up to 95 per cent. Therefore, gelled instruments should be higher in wattage. They are ideal for lighting backgrounds such as the cyclorama.

Lighting for News Programs

Most news broadcasts have an anchor or a pair of anchors sitting at a desk facing the cameras. Many TV studios also have an area behind the anchor painted chroma key green or blue. It is necessary to light talent with fresnel spots so that it does not spill onto the chroma key areas in the background. To light these areas, scoops are used. The intensity of the lights for the chroma key is kept lower than the foreground.

LIGHTING FOR INTERVIEW PROGRAMS

Interview programs with more than two participants tend to be a bit difficult to light. For a three person interview, each person is assigned a separate key, fill, and back light. The lighting is for the subject, not for the chair the subject is sitting in. This tends to be a problem because the subject will be facing other talent and if the lighting was set up incorrectly, the lights will be off the marks.

SOUND

Apart from the video, audio signal accompanies television signal. This may be mono or stereo. Sounds are actually a series of compressions and decompressions of the air around us. The changing air pressure causes the eardrums to vibrate. High frequency sounds, such as a football referee's whistle cause the eardrums to vibrate more rapidly than lower frequency sounds such as a bass drum. The brain interprets the strength and speed of these vibrations. The variations in pitch and loudness are commonly represented graphically as sound waves Fig.2.14. Sound consists of analog (continuous) signals. The varying heights and depths of the sound wave pictured below represent the amplitude (loudness). The frequency of a sound wave determines the pitch or tone of the sound. Frequency is measured in cycles per second, or Hertz (Hz). The lowest pitched sounds an average human can hear are about 20 Hz, and the highest are in the 20,000 Hz range.

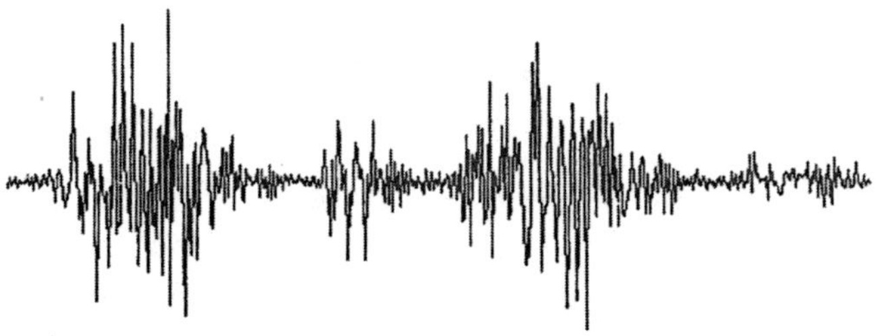

Fig. 2.14. A Typical Sound Wave.

CHARACTERIZING AUDIO SYSTEMS

LOUDNESS

Sound loudness is measured in decibels (dBs). The term decibel, dBm (for the milliwatt reference level) is a unit of electrical power. In audio production, dBm reflects levels of electrical power going through various pieces of audio equipment. 0dBm (generally just designated dB on a VU meter) refers to the maximum desirable sound level. The ideal range for meter is just below the red areas. Fig. 2.15.

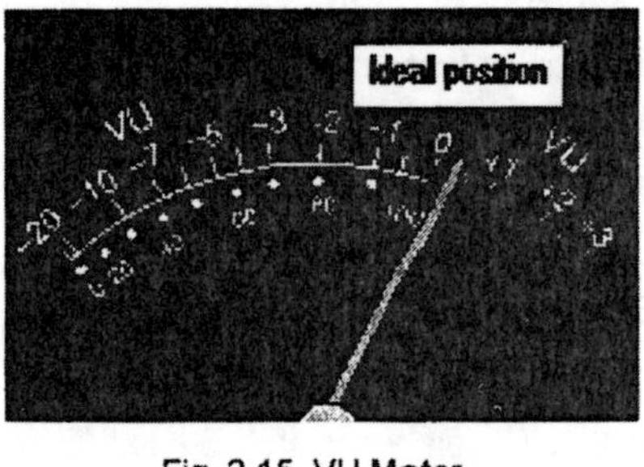

Fig. 2.15. VU Meter.

A graphic equalizer allows specific bands of frequencies to be individually adjusted for loudness.

FREQUENCY RESPONSE

Frequency relates to the basic pitch of a sound. Frequency is measured in Hertz (Hz) or cycles per second (CPS). The audio frequency range is from 20-20,000 Hz. Since both ends of the 20-20,000 Hz range represent rather extreme limits, the more common range used for sound /television production is from 50 to 15,000 Hz.

Frequency response characterizes the amplitude deviation in a signal over a given frequency range. Frequency response is typically measured by connecting an audio oscillator to the input of the system and measuring the input voltage as a function of frequency. Simultaneously, the output voltage is measured as a function of frequency. The ratio of the two (output over input) is the gain of the amplifier as a function of frequency.

SIGNAL TO NOISE RATIO

Signal-to-noise ratio incorporates all noise and pick-up sources. This includes power supply, ground loops, clock leakage and so on. The basic idea in measuring signal-to-noise is to make a recording at maximum modulation followed by a recording at no modulation and then calculate the ratio of the two. For example, in audio tape recording, the first recording is typically made at 400 Hz with 100 per cent modulation, and the second recording is made with no input. The difference between the two (in dB) is the signal-to-noise ratio. An alternative measurement technique, is to measure the signal-to-noise ratio as a function of frequency. For a digital recording, using a test disk with all zeros and measuring the output signal as a function of frequency performs the signal-to-noise measurement.

TOTAL HARMONIC DISTORTION

Audio systems often distort signals by adding harmonics to the primary signals. For example, if a 1 KHz sine wave is input into an audio system, a 2 KHz (2nd harmonic), 3 KHz (third harmonic), and so on may be produced. Inputting a pure sine wave at the input and then measuring the amplitudes of the resulting fundamental and harmonics measure is called the Total Harmonic Distortion (THD) Fig. 2.16.

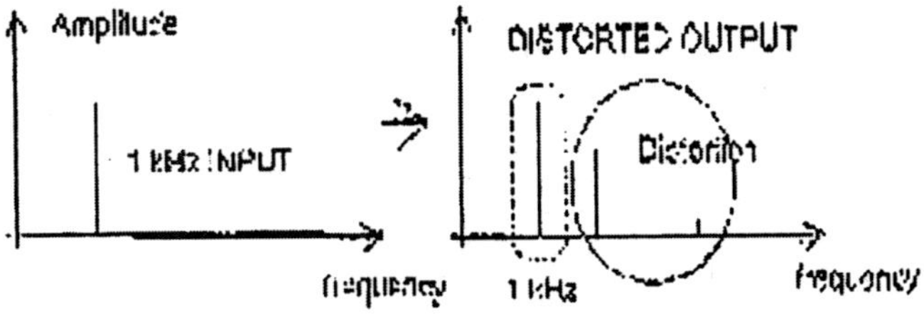

Fig. 2.16. Harmonic Distortion.

INTERMODULATION DISTORTION

When two frequencies are fed into an audio system, a nonlinear transfer function produces harmonics of both frequencies, sums and differences of both frequencies, and harmonics of the sums and differences, and sums and differences of the harmonics. In general, the harmonic distortion is separated from the intermodulation distortion. Thus, in an intermodulation measurement, only the sum and difference terms are taken. The Intermodulation Distortion (IMD) is measured by inputting two pure sine waves at the frequencies f1 and f2 and then measuring the amplitudes of the resulting fundamental and harmonics.

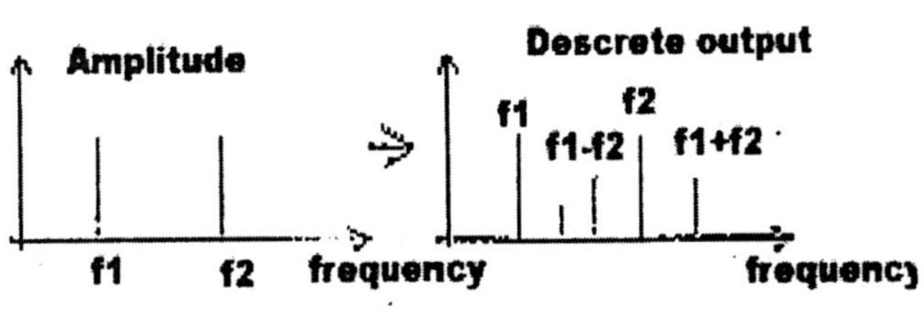

Fig. 2.17. Intermodulation Distortion.

MAGNETIC RECORDING

The simplest recording system consists of a ring-shaped electromagnet with a ferrous core mounted over a ferromagnetic surface travelling at a velocity. Since the core of the electromagnet is ferrous, the flux preferentially travels through the core. Thus, the core is deliberately broken at an air gap. In the air gap, the flux creates a fringing field that extends some distance from the core Fig. 2.18.

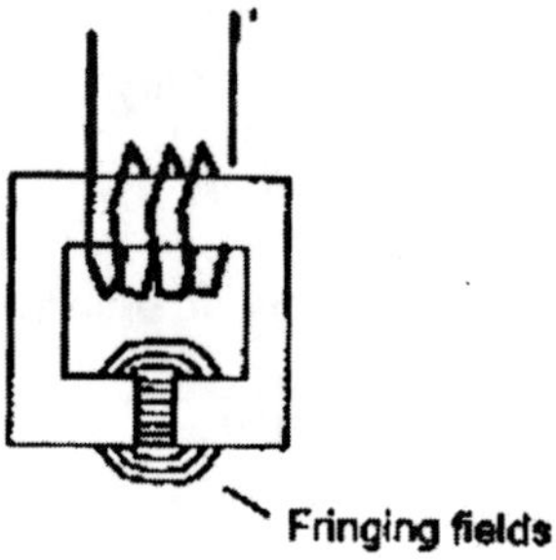

Fig. 2.18. Fringing Fields Across the Core.

The signal current through the electromagnet generates a fringing magnetic field H. The fringing magnetic field H then creates a permanent magnetization on the ferromagnetic surface (Fig. 2.19).

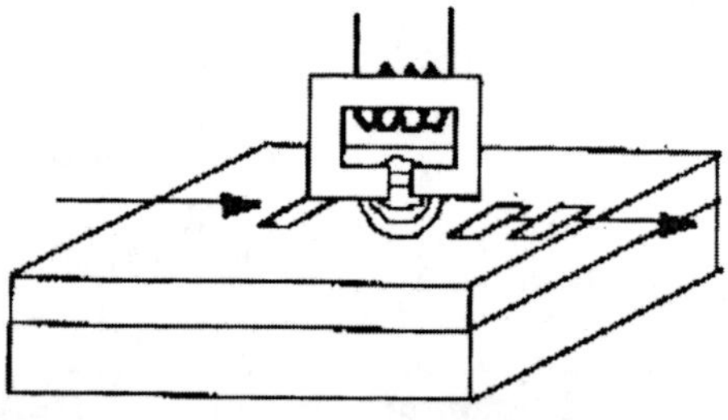

Fig. 2.19. Permanent Magnetization on the Ferromagnetic Surface.

Thus, the ferromagnetic surface becomes permanently magnetic. The magnetic particles in the surface act like little bar magnets themselves and create their own fringing magnetic field H above the ferromagnetic surface Fig. 2.20.

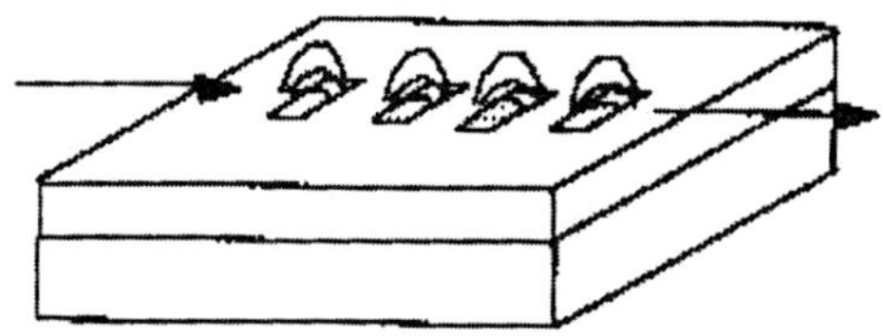

Fig. 2.20. Little Bar Magnets with Fringing Magnetic Field.

An analog signal passing into the electromagnet creates an analog variation in H Fig. 2.21. If the surface is moving in time, the analog variation of the signal is translated into a magnetic permanent variation on the surface in space.

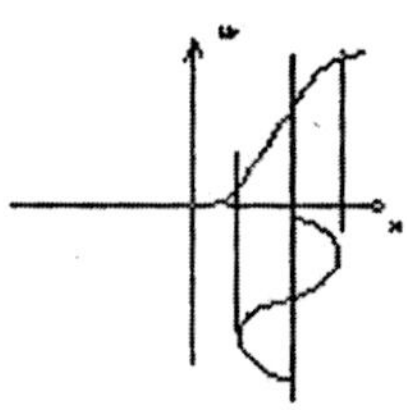

Fig. 2.21. Analog Variation in H.

Thus, a very important number in magnetic recording is the *spatial variation*, which corresponds to a *signal frequency*. If an incoming analog signal has a frequency f, then the characteristic wavelength of the pattern written on the magnetic media has a wavelength given by

$$\text{wevelength} = \frac{\text{velocity}}{f} = \frac{\text{m/sec}}{\text{cycle/sec}} = \text{m/cycle}$$

READING

The simplest reading system consists of the same head that was used to write the data on the ferromagnetic surface travelling at the same velocity. The head now passes over the fringing magnetic field H above the ferromagnetic surface Fig. 2.22. The voltage then induced in the electromagnet is proportional to the spatial (i.e. time, since the tape is moving) derivative of the magnetic field created by the permanent magnetization in the material.

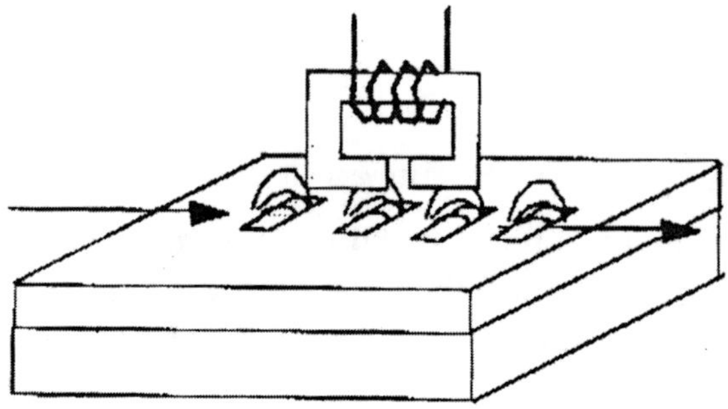

Fig. 2.22. Reading the Tape.

A TYPICAL TAPE SYSTEM

A typical audio tape recorder has three heads: erase, record and reproduce. During recording, the tape first moves over the erase head, where all trace of earlier programming is removed. The erase head also leaves the head in the maximum random condition for lowest possible noise. Therefore, the erase head gap length is large to provide a strong erasing field. The erase head is also fed with an A.C. signal in order to minimize noise. Next the tape moves over the record head. The record head is usually fabricated with a gap size as per the order of the thickness of the magnetic coating. This provides a compromise between low and high frequency response. Further, the signal going into the record head is usually altered in two major ways from the original signal. First, an ac component is added to the signal to linearize the response. Secondly, the frequency spectrum is altered to compensate for losses that occur in the recording process for high frequency signals. This compensation process, called pre-equalization, provides the tape recorder a flat frequency response up to its maximum design frequency. Finally, the tape moves over the reproduce head. The reproduce head is usually fabricated with a small gap in order to maximize high frequency response. The inductive pick-up generates a signal, which is the derivative of the flux, so the signal must be integrated. Additional equalization, called post-equalization, is added to compensate for losses that occur in the reading process for short high frequency signals.

THE MEDIA

A variety of magnetic media have been used over the years. Very early recorders used ferrous wire. However, most modern magnetic media use a thin layer of ferromagnetic material supported by a non-magnetic substrate. The magnetic layer is formed of magnetic particles (such as gamma ferric oxide) in a polymer matrix. Alternatively, the layer is a vacuum deposited metal or oxide film.

AUDIO CONTROL DEVICES

Audio control devices are used to maintain proper audio levels. Few of these are described below:

AUTOMATIC GAIN CONTROL (AGC) CIRCUIT

AGC circuit maintains the audio level through a feedback circuit. Thus, the gain remains constant.

LIMITER

Limiter keeps the audio level from exceeding a set level, generally 0dB.

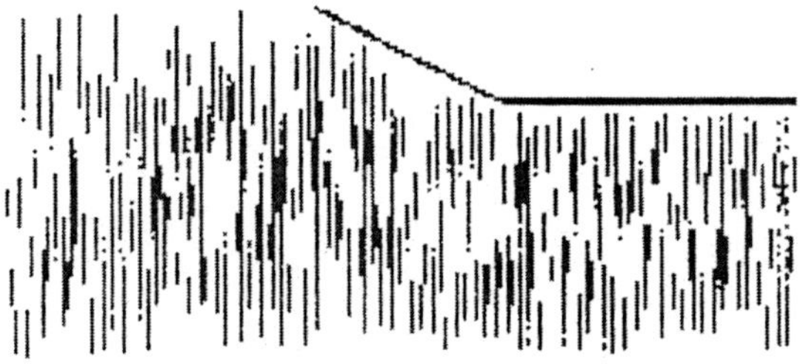

Fig. 2.23. Level Control by a Limiter.

DIGITAL AUDIO

Computers digitize sound waves—convert them to numbers by taking "snapshots" of that wave at regular intervals. The number of snapshots the computer saves for each second of audio is known as the sampling rate of the sound. Higher sampling rates and quantization levels produce more realistic sound. However, higher sampling rate, say 44.1 KHz for CD quality, requires huge storage space. Compression schemes are used to reduce the audio.

44,100 [samples per second] X 2 [For 16 bit (8 bits = 1 byte)]=88,200 bytes per second=88,200 X 2 [stereo (2 channels)] = 176,400 bytes per second

Thus, for a sampling rate of 44,100, a 16 bit stereo will need storage space of 176,400 bytes per second.

There are a number of ways to digitize audio signals. The most conceptually straightforward of the various techniques is pulse amplitude modulation. In pulse amplitude modulation, one pulse occurs at each sample interval, and the amplitude of the waveform is simply a digital signal corresponding to the analog amplitude.

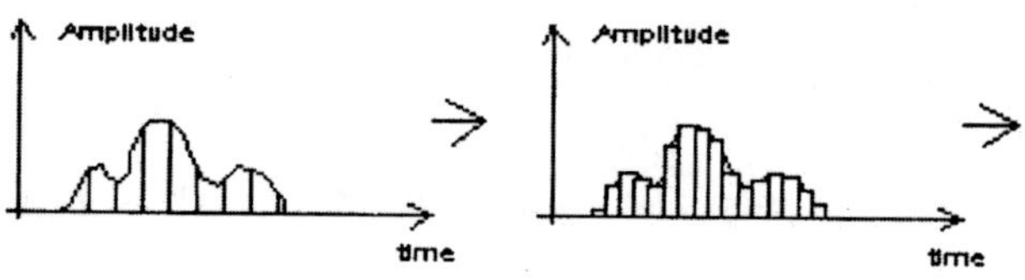

Fig. 2.24. Pulse Amplitude Modulation.

In pulse position modulation, the width of the pulses remains constant, but their position in time carries the digital information Fig. 2.25.

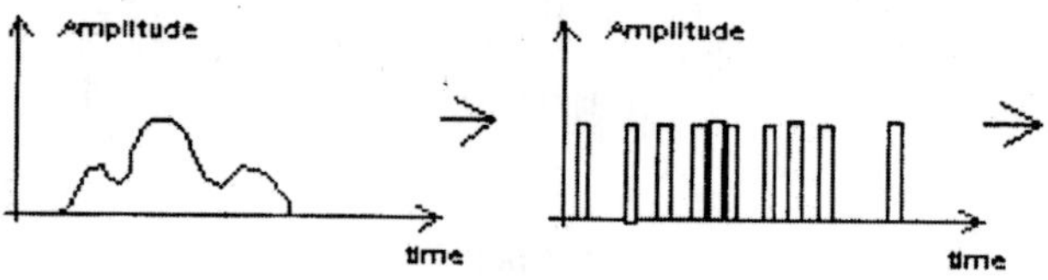

Fig. 2.25. Pulse Position Modulation.

In pulse width modulation, one pulse occurs at each sample interval, but the width contains the information Fig. 2.26.

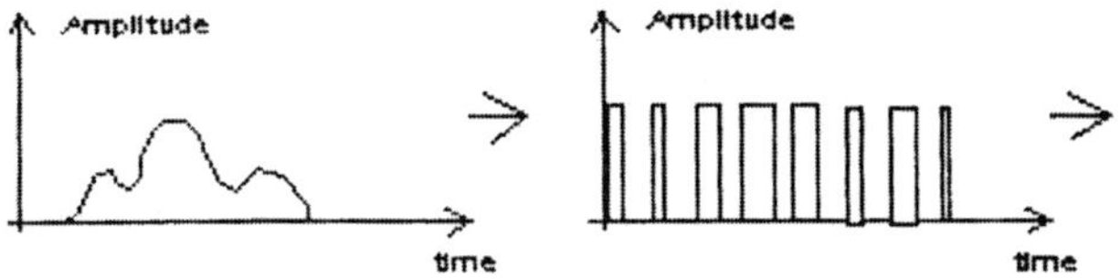

Fig. 2.26. Pulse Width Modulation

In pulse number modulation, the number of pulses per sample interval is the information Fig. 2.27.

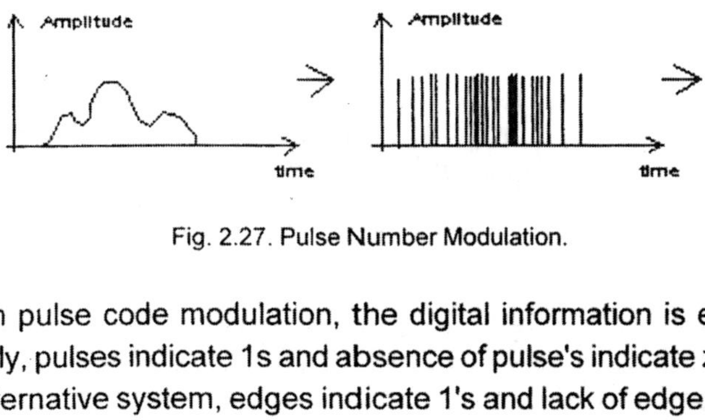

Fig. 2.27. Pulse Number Modulation.

In pulse code modulation, the digital information is encoded serially, pulses indicate 1s and absence of pulse's indicate zeros. In an alternative system, edges indicate 1's and lack of edge indicate zeros.

COMPANDING

In general, most of the important information is in the lower levels of amplitude. Thus, faithful reproduction of the volume levels of quieter portions of the sound waves is more critical to making the resulting audio seem more realistic to the ears than exact reproduction of the louder portions. The audio compression codecs take advantage of this fact by compressing the data in the higher levels and expanding the data in the lower in order to preserve more of the quality of those parts of sound which are most vital. This is known as companding (compacting/expanding) Fig. 2.28. An audio expander increases the dynamic range of audio that has been overly processed. An expander can restore the audio to its normal range and in the process reduce noticeable background noise. There are two separate standards for companding. The U.S./Japanese standard is known as µ-Law, and the European standard is A-Law. Both accomplish the same objective, but the methods differ slightly.

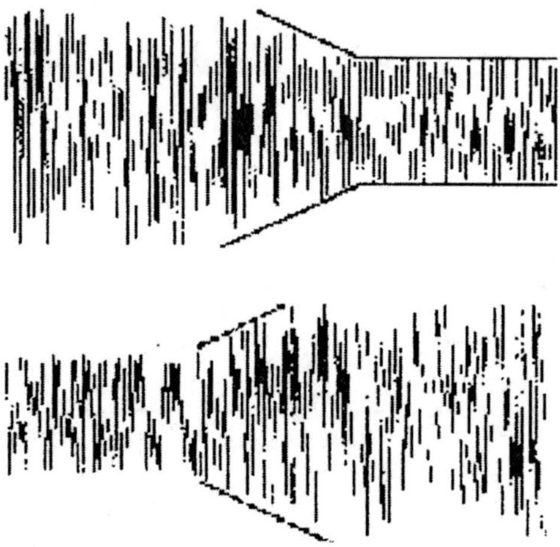

Fig. 2.28. Audio Compander and Expander.

PERCEPTUAL CODECS

Most of the newer audio codecs take advantage of the fact that many of the sounds present at any given time are never heard by the ear, even though they lie within the frequency spectrum of human hearing. They are overpowered by other sounds that are occurring simultaneously or almost simultaneously. Although the vibrations from these sounds reach the eardrums, the brain discards their input and never processes them, so in effect they are not heard at all. This phenomenon is known as psychoacoustic masking. The codecs, which take advantage of this masking, discard all such data when they compress the sound, resulting in significant savings in audio size with little or no loss in apparent sound quality. Such codecs are known as perceptual codecs, because they attempt to reconstruct only those parts of sound that human mind actually perceives.

ACOUSTICS

The primary function of a recording studio is to provide adequate isolation between microphones while allowing the players to hear each other as well as possible. The sound of one instrument or section must not trash the sound of another. Acoustical isolation between instruments is a function of the degree to which the sound of one dies away before getting to the next. When the die off is inadequate, the sound of one instrument falls through the mike of the next and trashes it. If it is excessive, the musicians do not hear each other properly, which makes group playing difficult and ruins section sound. The amount of acoustical attenuation for a given instrument in a room depends on the room's global characteristics. As with any radiated field, sound pressure levels diminish as the square of the distance from the source. The equation holds for any distance in a perfectly dead room or out of doors.

Studio isolation depends on several things viz. the directionality of instruments / mikes, room treatment, etc. If there is sound bouncing around in a room, the bounce at some point becomes as loud as the sound source that created it. That point is called the critical distance Dc. Dc of a sound source indicates the distance in a room to which the sound travels before it increases to constant volume. Dc varies with a number of factors including the directionality of the source, reflectivity of the walls and the area of wall surface on which the sound hits. As rhythm section instruments have about the same directionality as voice, a general treatment can be made for rhythm and vocal isolation. Control wall is used to supply a live surface to the musicians. It is made more effective by retreating the wall for maximum reflection, moving the control room window to the upright position, and installing a reflector above the window angled to bounce even more sound back to the rhythm section. The combination of flat and dispersed reflections has been shown to be optimum for critical listening. The treatment needs to yield 26 db or die off before the source gains constant volume. Sound dies off by 6 db per distance doubling until the Dc is reached, with the last figure of a 3 db point. Therefore, assuming a mike to mouth distance of six inches, 1ft—6db, 2ft—12db; 4ft—18db; 8ft—24db, and as the last figure adds only 3db, a voice Dc of 16 feet adds 3 db more for 27 db of acoustical loss at that figure.

Flat, straight walls are acoustically unacceptable. That is partly because sound reflects off such walls as a flat smack and partly because it bounces strongly. The obvious remedy lies in installing absorptive treatment in patches.

Both objectives are accomplished by hanging live-sided boxes filled with Fiberglas on the walls. The boxes start high enough off the floor to clear chairs and other clutter leaned against the wall, and generally top out at eight or nine feet above the floor. Floor to ceiling treatment is unnecessary, as mikes rarely point up.

A box with reflective sides acts as a disperser. These are normally spaced at three to six feet apart, leaving the walls reflective between them. This presents a combination of dispersion, absorption and reflection to both the musicians and the mikes, and presents an optimum listening environment in the studio.

The back wall is treated in the same way in small rooms, although it is best to leave the back as live as possible, as reflections from it give the players a sense of being in a room rather than working with their backs to a vacuum.

In cases where a great deal of absorption is needed, the wall area above eight feet and below about two, can be totally treated without ruining the generally live sound of the room, as the ear only needs a little encouragement to think it is in a normal environment.

The acceptable room proportions are of up to two to one, and the vast majority of ceilings are well over that. Since a non-reflective surface generates no mode structure, dead ceilings are the norm in most studios. The ceiling is kept almost entirely dead, because it is never high enough to establish a decent mode structure.

The ceiling also presents the largest area available for treatment, especially as it can be totally absorbent without making a room sound dead. Unless a ceiling is extremely low, the ear ignores it, preferring to take its cues more or less horizontally. It is critical that the ceiling treatment be acoustically flat in its absorption. Given an ordinary grid hung 16 inches below the structural ceiling, flat response can be accomplished with 1-1/2 inch Fiberglas ceiling panels or with thinner panels and a Fiberglas overlay.

The low-end absorption of the ceiling must extend far enough into the bass range to avoid the common fault of acoustical treatment that soaks out the top end of the room while leaving the low end live. That kind of treatment results in a muddy room with terrible isolation problems in terms of bass, floor tom, and bass drum.

The need for flat low-end response applies to all room treatment unless the studio has big windows or its walls are so flimsy as to transmit or absorb bass by vibrating to it. Even then, bass attenuation seldom exceeds 30%, leaving 70% to be supplied by other means. On balance it is more practical to install the general treatment in such way as to absorb uniformly from bass to cymbals.

Controlling high frequency reflections is easy, but bass absorption is largely a matter of absorber depth. It takes considerable thickness to get flat down to 60 Hz. Hung ceilings manage it with thin panels and the 16 inches between panels and the real ceiling, but a wall mounted absorber needs a minimum depth of 6 inches for Fiberglas (703) board, and a foot for glass wool.

RADIO FREQUENCIES

For convenience the radio frequencies are split up into areas called: M.F. (Medium Frequency), H.F. (High Frequencies), V.H.F. (Very High Frequency), U.H.F. (Ultra High Frequency) and Microwave. The features of these frequencies are given below:

MEDIUM FREQUENCY (MF)

A groundwave and a skywave characterize the MF band that includes AM broadcast (550 to 1550 kHz). There are no multipath problems and no terrain effects. Low cost hardware operating at medium power levels provides for a wide coverage area. The problem is man-made noise and atmospheric noise can degrade the signal. The other characteristics of MF are:

- From 300 KHz (0.3 MHz) to 3 MHz.
- Local communication.
- Very large antennas.
- Commonly uses S.S.B. and C.W.
- RTTY (Radio Teletype), SSTV (Slow Scan Television) and Packet (at 300 Baud) are all used on the M F band.

HIGH FREQUENCY (HF)

The high frequency extending from 3 MHz to 30 MHz is used for short wave communication. The coverage is much larger as compared to MF but is subject to ionosphere conditions. The other characteristics are:

- "Around the World" communications.
- Large antennas required
- Commonly uses S.S.B. and C.W. sometimes A M is still used.
- RTTY (Radio Teletype), AMTOR, PACTOR, SSTV (Slow Scan Television) and Packet (at 300 Baud) are all used on the H F bands.
- In general, the lower frequencies (less MHz) are good after dark and the higher frequencies are good during the day.

VERY HIGH FREQUENCY (VHF)

The VHF band, which includes FM broadcast and VHF television provides the widest area coverage at reasonable power levels, has good building penetration, and moderate operating costs. Its drawbacks are moderate terrain effects and multipath. The other characteristics are:

- From 30 MHz to 300 MHz.

- "Line of Sight".
- Small antennas, higher up is better.
- Commonly used for local communications with FM portable or mobile radios.
- Long distance communications are possible under some conditions. S.S.B. or (Morse Code) are often used under these circumstances.
- Packet Radio commonly uses FM radios (the same as voice communications) at 1200 Baud for computer communications.
- In few countries, there are three bands available for amateur radio use in the VHF area. 50 MHz to 54 MHz, 144 MHz to 148 MHz and 220 MHz to 225 MHz.
- 88-108 MHz is reserved for FM broadcast. Each FM network occupies 2.2 MHz of spectrum.

FM broadcasting stations can be operated in the very-high-frequency bands at which AM interference is frequently severe; commercial FM radio stations are assigned frequencies between 88 and 108 MHz. The range of transmission on these bands is limited so that stations operating on the same frequency can be located within a few hundred miles of one another without mutual interference.

ULTRA HIGH FREQUENCY (UHF)

The UHF band provides wide area coverage but at high power levels. It is characterized by significant terrain effects, low building penetration rates, the need for high gain antennas, and high operating costs. The other characteristics are:

- From 300 MHz to 3000 MHz (3GHz)
- "Very Line of Sight".
- Very small antennas (a few inches).
- Increasingly used for local communications with FM portable or mobile radios.
- Long distance communications (including Moon Bounce are possible under some conditions. S.S.B. or C.W. (Morse Code)

are often used.

Packet Radio commonly uses modified FM radios at 1200 Baud and above for computer communications.

MICROWAVE

The microwave band requires very high power levels for wide area coverage as well as high-gain antennas. Microwave also has serious terrain issues, limited reception inside buildings, and high operating costs. It does provide a low noise signal environment. The other characteristics are:

- Super High Frequency.
- From 3 GHz (3000 MHz) to 30 GHz (30000 MHz).
- Extremely Line of Sight.
- Radio waves act more like light waves.
- Very small antennas (a few inches).

Long distance communications are possible under some conditions by reflecting the radio waves off objects.

CONCLUSION

In this chapter, the basic broadcasting system including type of transmissions, working of TV system, video basics, various TV standards, lighting techniques, audio system, acoustics and radio frequency spectrum have been described.

REFERENCES

1. Gordon J. Harold, "Digital Radio World-Wide: Turning the Drean into Reality", *Technical Review*; Asia-Pacific Broadcasting Union, No.161, pp. 17-25, Nov.-Dec. 1995.

2. John Andrews, "Audio at IBC 95", *International Broadcast Engineer*, pp. 16-17, Nov. 1995.

3

RECORDING AND STUDIO TECHNIQUES

In the first chapter, the historical development of the broadcast technology and few milestones thereof, were described. In the second chapter, the broadcast basics including audio and video technology and general broadcasting system were discussed. In this chapter, the broadcast technology components, the production processes and sound recording techniques of sound /TV broadcasting are being described. The recording and studio technique of radio communication encompasses a number of technology and processes. The technologies involved are acoustic engineering, air conditioning, sound engineering, video engineering, lighting and power engineering, etc. The processes are pre-production, post-production and distribution. These processes have undergone a complete change from the early days of analog recording on LPs and Tapes to the present era of digital system using DAT/DVT (Digital Audio/Video Tape), CDs (Compact Disks) etc. In this part, studio techniques including production processes, earliest recording methods, microphone and camera characteristics, recording devices, mixing consoles and acoustic treatment have been described.

The history of broadcast development starts from the Marconi's experiment, which used a key (Morse key) to broadcast its first signal across the Atlantic Ocean[1]. The first voice transmission was done on December 24, 1906 from Brant Rock MA, which included a female voice singing a Christmas carol and Fessenden doing a speech and a violin solo[2]. The tradition of live broadcast continued until mid of the current century when the recording processes started. The post-production at studios matured in 60s to 80s. The last decade saw the onset of digital technology, non-linear editing, virtual studios, etc.

Thus it would be seen that, while the early broadcasts were live, soon the importance of recorded broadcasts was realized. The recording technique provided flexibility of editing, mixing and re-usability. As such the recording and studio production is the essential component of modern broadcasts. The broadcast process at a Radio/TV station, essentially consists of initial recording/shooting, post-production and distribution Fig. 3.1. Initial production involves actual audio /video recording in a studio or at the place of activity using microphones, cameras and allied equipment like amplifiers etc. The post-production involves editing, dubbing, mixing of the effects/ scenes, and finally creating a record ready for distribution. A variety of new equipment / techniques are available today for post-production viz. virtual studios, non-linear editing, etc. The final distribution involves a transmitter and the channel, discussed above.

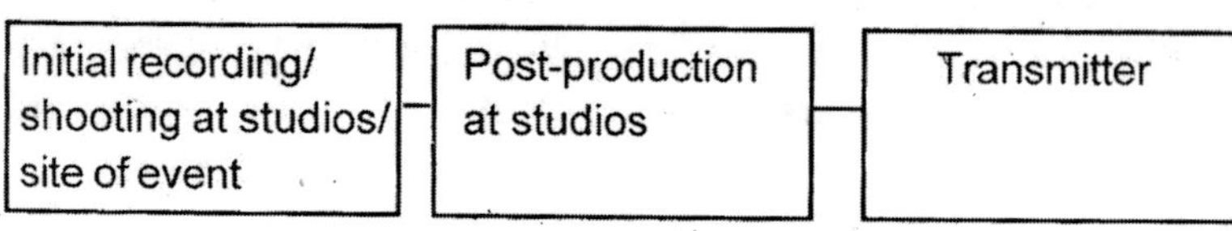

Fig.3.1: Broadcast Processes.

A variety of technologies are involved in the processes like air conditioning to control the temperature and humidity, acoustic engineering to control the echo and reverberation, lighting control for proper operation of cameras, etc. A brief description of acoustics and lighting was given in Chapter 2. Air conditioning, which is a subject in it, shall not be decried in this book.

EARLIEST RECORDINGS

MORSE EMBOSSER (CHAIN-DRIVEN MORSE RECORDER)

Fig.3.2. Morse embosser (chain-driven Morse recorder).

The earliest recorder was the embosser or the chain driven Morse recorder Fig. 3.2 developed by Samuel F.B. Morse (1791-1872) who invented the electric telegraph named after him. These embossers used dry pin for writing. In February 1852, seven sets of such Morse embossers, developed by Messrs. Carl Geiger, Stuttgart, were first used for instructional purposes and then employed by the Federal Telegraph Workshop in Berne as models for manufacturing their own Morse apparatus. The Federal Department of Posts and Public Works, Switzerland deployed these on 5 December 1852. After 1856, the clumsy driving weights, suspended on chains, gradually gave way to a spring movement. From 1863, the embosser writing with a dry pin, was replaced by the ink writer, which greatly facilitated reading of the dash-dot Morse code on the paper tape. The last Morse instrument of the Swiss PTT was put out of service in 1953[3].

VOICE RECORDING

Edison did the first recording of a human voice ("Mary had a little lamb") on 6th December 1877 using a tinfoil cylinder phonograph

Fig. 3.3. The phonograph had a capacity for recording of 2-3 minutes. Edison later introduced wax cylinders, but was unable to mass-produce copies. It was the year 1888, when Emile Berliner patented the "Gramophone" using a flat 7 inch disk with lateral cut grooves on one side only, manually rotated at 70 rpm with 2 minute capacity (Fig. 3.4). He was the first to mass-produce rubber vulcanite copies from zinc master disk. There were a number of other modifications and improvement by Edison, Odeon, Bing Crosby (used condenser mike developed in 1922 by Bell), Peter Goldmark, etc. These and several other efforts lead to the development of 12 inch, 331/3 rpm micro-groove LP vinylite record with 23 minute per side capacity in 1947 by Philco, 7 inch 45 rpm micro-groove extended play vinylite record and player developed by RCA Victor in 1949, Nagra portable audio tape recorders by Stefan Kudelski in 1951, the first experimental 12 head VTR at 100 ips, by Ampex team led by Charles Ginsburg in 1951, etc[4]

Fig. 3.3 Edison's Phonograph in Scientific American Dec. 22,1877

Fig. 3.4 Berliner's 1st model of the Gramophone.

'PHONOVISION' : THE WORLD'S FIRST RECORDINGS OF TELEVISION

Scottish inventor, John Logie Baird, succeeded in becoming the first to record television on wax discs, in 1927[5,6,7,8]. With only 30 lines per picture (television frame), the highest frequency present was low enough to be audible. The video signal could therefore be recorded as an audio signal onto the disc. The 30 line images did not take up the entire "screen", but were in fact 6 cm high and 2 cm wide. Instead of black and white, they were black and red due to the colour of the neon gas in the lamp. Baird was one of the true pioneers of television. He successfully demonstrated the possibilities of the Nipkow system of mechanical television.

The BBC started broadcasting television on the Baird 30-line system experimentally from 1929. The first simultaneous sound and vision play was broadcast in 1930 followed by regular programmes of entertainment from 1932. Transmissions on the BBC continued until 11th September 1935 as developments in fully electronic television led to the demise of the 30 line service. The BBC replaced it with the world's first regular high-resolution (405 lines per picture) television service in 1936.

THE FIRST TV RECORDING STUDIO

Fig. 3.5. World's first TV recording
Copyright ©DFMcLean 1996.

The world's first TV recording studio is seen in the Fig. 3.5. A record turntable was linked to the drive shaft of a 'camera', which is hidden behind a wooden paneled wall. As the horizontal drive shaft (disappearing into the wall on the left) turns, the worm gear transfers the rotation to the turntable. The drive shaft was connected directly to a Nipkow scanning disc capable of scanning one 30-line frame on each rotation. Baird used large Nipkow discs to carry the large aperture lenses needed to gather the light[9].

STUDIO CHAIN

The recording is done either in the studios or outside at the The place of event. The outside recording is also known as OB (Outside Broadcast). A transducer known as microphone, which is responsible for converting sound waves into electrical signals, picks up the sound. Cameras of different types are used for capturing the picture or the event in video form. All videos and films are created from the same elements like close-up shots, titles, music and narration. When skillfully combined, these visual and audio elements become a powerful tapestry much greater than the sum of its parts. The levels are raised to the desired level by using amplifiers. Mixers/ Faders are used to mix/select the output from various channels and thereafter video/audio tape recorders are used for recording it on the media. In the succeeding paras, various pickup devices and the equipment used in the chain are described.

MICROPHONES AND THEIR CHARACTERISTICS

Microphones capture sound and transform it into electrical impulses. The generating element of a microphone is the material that converts the acoustic energy into electrical energy. Microphones are classified according to the type of element used in its construction. There are three types:

Dynamic Microphones

In a dynamic microphone (also called a moving-coil microphone) Fig. 3.6 sound waves hit a diaphragm attached to a coil of fine wire. The coil is suspended in the magnetic field of a permanent magnet. When sound waves hit the diaphragm it moves the coil of wire within the magnetic field. As a result, a small electrical current is generated which is amplified thousands of times.

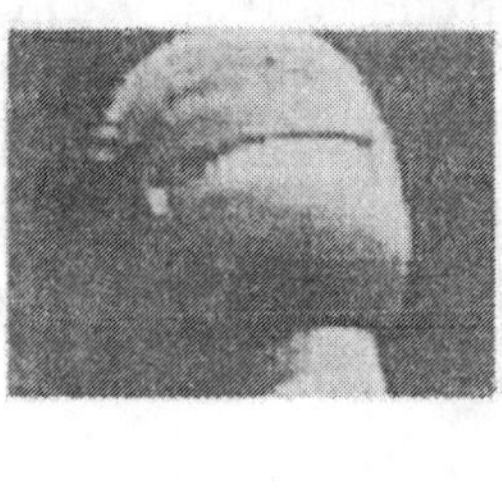

Fig. 3.6. Dynamic Microphone.

dynamic mike is considered the most rugged professional microphone. This type of mike is a good choice for sound recording in the studios, field production and also electronic news gathering (ENG). Dynamic microphones are also very useful for recording loud and sudden sound effects, such as crashes and explosions. Not only are these microphones virtually impervious to damage from high noise levels, but they also tend to compress or dampen the audio in such a way as to make these sounds easier for the recording electronics to handle.

Condenser Microphones

Condenser microphones (also called capacitor or electric condenser mike) Fig. 3.7 work on the principle of an electric condenser or capacitor. An ultra-thin metal diaphragm is stretched tightly above a piece of flat metal or ceramic. Sound waves hitting the diaphragm cause fluctuations in an electrical charge, which is maintained using a power source. A pre-amplifier (pre-amp) amplifies the varying electric charge. The pre-amp is located within the microphone housing or in a outboard electronic pack.

Since they require a pre-amp, most condenser mikes require a source of power, either from an AC power supply or from batteries. AC power supply is often built into an audio mixer or audio board. This is referred as phantom power supply. The mike cord apart from delivering the signal from the mike to the mixer carries power from the mixer to the pre-amp of the condenser mike.

Sometimes, two miniature condenser mikes are used together. If one mike goes out, the other can immediately be switched on. This double microphone technique is called dual redundancy.

Small size, optimum sensitivity and the best quality characterize condenser mike.

Ribbon Microphones

Ribbon microphones use a thin metal strip as the vibrator. Like dynamic microphones, the electrical pulses generated by the movement of the mikes are picked up. Due to the nature of its design, ribbon microphones are extremely fragile and usually used in sound recording studios. Their quality is superior and best suited for announcers and singers.

Fig. 3.7 Condenser Microphone.

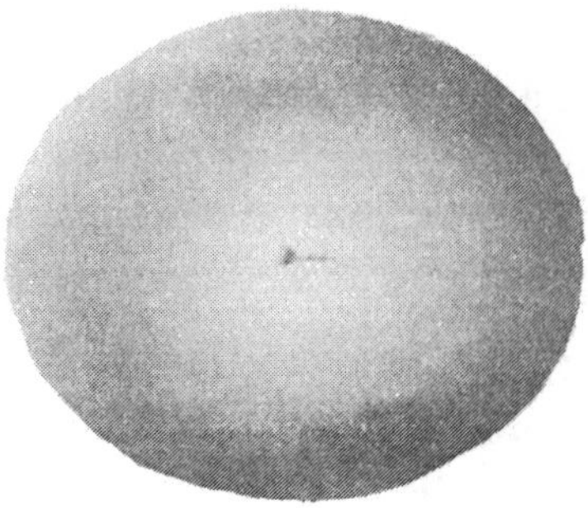

Fig. 3.8 Omni-directional Plattern.

DIRECTIONAL CHARACTERISTICS

Microphones are also classified according to the shape of the area of sensitivity around the microphone. Directional microphones are designed to be more sensitive in some directions than others. A polar diagram provides this description. The microphone faces the 0° axis. The 180° axis is located at the rear of the mike. Microphones have three basic directional patterns:

- Omni-directional
- Bi-directional
- Unidirectional

Omni-directional Mike

Omni-directional mikes (also called non-directional mike) are equally sensitive to sounds coming from all directions. Fig.3.8. It is very flexible and an omni can be placed almost anywhere in most situations and pick-up usable sound. Omnis are made for various purposes. One kind of omni is the microphone built into the camera. This microphone allows one person to handle both the audio and video. Another kind is called the lavalier, a small mike that clips to a lapel or other part of the wearer's clothing. It can be hidden and leaves the performer's hands free.

The Omni-directional make's characteristics have definite advantages in radio, in recording of discussions where several people stand or sit around a single microphone. However, in video production it is more desirable to use some form of directional mike to reduce unwanted sounds (behind-the-camera noise, ambient on-location noise, etc.) while maximizing sound coming from talent.

Bi-Directional Mikes

In a bi-directional (also called figure eight) mike, sensitivity pattern of the mike is polar and primarily the mike is responsive to sounds from two directions. The mike with this sound pattern picks up sounds directly in front (0° axis) and directly behind it (180° axis).

This is commonly used in radio for interviews or drama where people sit facing each other at a table. The bi-directional sensitivity pattern has limited use in television, since most talent face the camera when performing. As a result the microphone would be subject to pickup of camera noises along with the sound from the talent.

Un-Directional Mikes

Unidirectional mikes are sensitive to sounds coming primarily from one direction. There are four categories of un-directional microphones:

- Cardioid
- Supercardioid
- Hypercardioid
- Parabolic

These classifications are based on mike's pick-up pattern ("angle of view") Fig. 3.9.

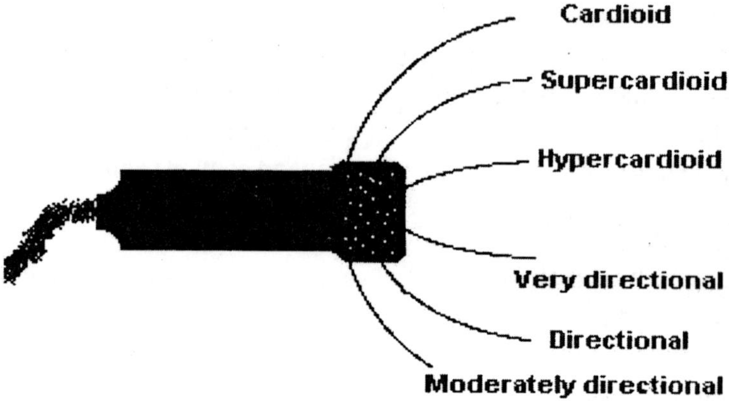

Fig. 3.9. Pick-up pattern of different mikes.

Cardioid

The cardioid (pronounced car-dee-oid) mikes are sensitive to sounds over a wide range in front of the mike, but relatively insensitive to sounds coming from behind the mike. The name "cardioid" comes from the shape of the pickup pattern Fig. 3.10 on a directional microphone which vaguely resembles an upside-down heart.

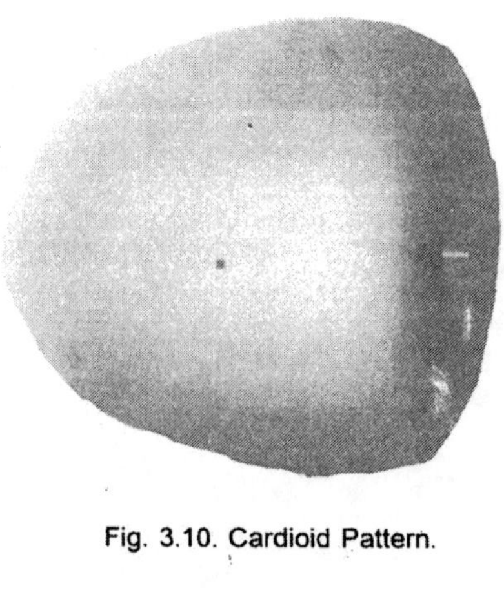

Fig. 3.10. Cardioid Pattern.

This pattern is useful for picking up a choir in a studio. However, the width of a cardioid pattern in front is too great for most TV applications. When placed at a distance of about eight feet or more from a speaker, it tends to pick-up unwanted, surrounding sounds, including reverberations from walls.

Supercardioid

The supercardioid mikes, as shown in Fig. 3.9, are more directional than the cardioid mikes. The sensitivity pattern of supercardioid mikes is polar like human ears. When the mike is pointed towards a sound source, interfering (off-axis) sounds are rejected. Shotgun mikes, widely used in on-location video production, are a type of supercardioid mike. The directional pattern of this mike provides good pick-up when used at a distance of 8-15 feet from the talker.

Hypercardioid and Ultra-directional

Hypercardioid or ultra-directional, as the name suggests are highly sensitive to one direction Fig. 3.11. Some highly directional shotgun mikes are included in the hypercardioid category.

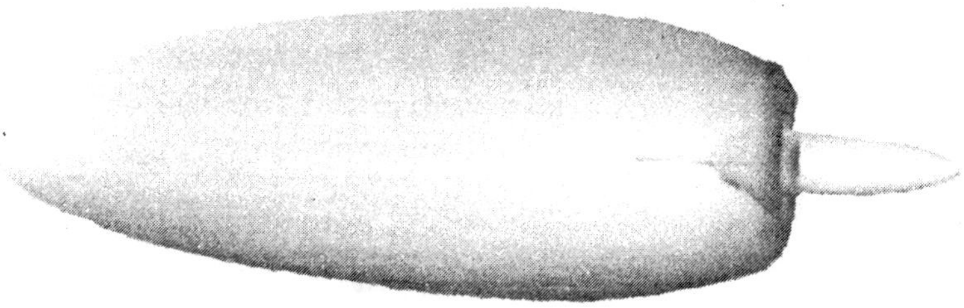

Fig. 3.11. Shotgun Pattern.

Due to the narrow angle of acceptance, off-axis sounds are largely rejected. As such, these mikes need to be accurately pointed toward sound sources. Regular adjustments have to be made if the talent moves.

Parabolic Mikes

In Parabolic mike, a parabolic reflector, 1 foot to 3 feet in diameter, is used Fig. 3.12 to create a polar pattern for the mike. Any type of basic uni-directional mike is used at the focal point of

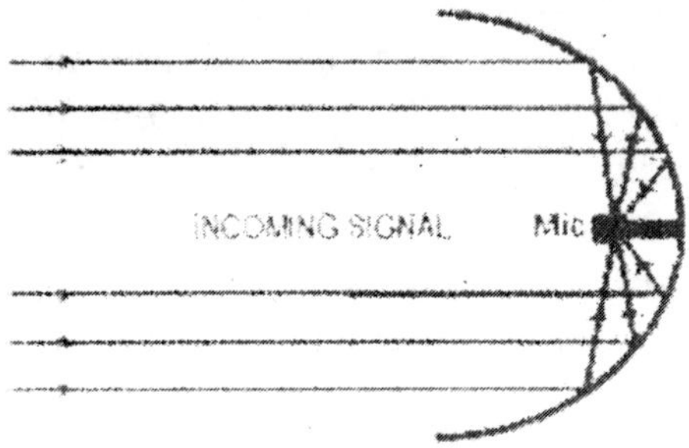

Fig. 3.12. Parabolic Mike.

the parabolic reflector. Because of the parabolic shape of the reflector, all the sounds travel along a very narrow angle of acceptance to the microphone. Therefore these are highly directional. The directional pattern is shown in Fig. 3.13.

Fig. 3.13. Directional Pattern of a Parabolic Mike.

Parabolic microphones are capable of picking up sound at distances of 300 feet or more. However, these mikes are not a practical choice for general field production work, albeit, they are often used in sports.

Hand-Held Microphones

Hand-held mikes are designed to be used at a distance of about 8- to 16-inches, albeit, the performer, in high-noise situations further reduces this distance. These are often dynamic mikes for better handling of sound overloads associated with some performers. Speaking or singing directly into mike from close distances, creates unwanted sibilance (an exaggeration and distortion of high-frequency "S" sounds), pops from plosive sounds (words with initial "p's," and "b's") and an undesirable proximity effect (the exaggeration of low frequencies). As such pop-filters are built into many hand-held mikes. Some pop filters cause high frequency roll-off. But this is usually corrected by 2 or 3 dB's boost in the 10 to 15 KHz regions. A windscreen, which is a small cover, usually of foam rubber, is used over the mike to further reduce the effect of plosive speech sounds. In addition to reducing the effect of plosives, windscreens eliminate the effect of wind moving across the grille of the microphone. (Even a soft breeze can create a turbulence that can drown out a voice.) Further, it is recommended to tilt the mike at about a 30-degree angle and not to hold it perpendicular to the mouth.

An elaborate windscreen is used in field production work. These are usually made of open-cell urethane foam and are available in many colors (darker colors are best for video). A highly directional mike is housed inside a foam and fur-covered shell. Generally, this type of mike is attached to a "fish pole" and held above the talent. A big fish pole is called a boom. A boom is a special microphone stand that allows the mike to be hung over the subject(s) facing the camera. To keep the microphone out of the shot, these stands are adjusted for height and angle. In addition, the microphone is swiveled 300+ degrees to pick-up sound from any subject.

Hand-held mikes are most often used in EFP/ENG as well as in the studio for musical ensembles.

Headset Mikes

The headset mike is a un-directional dynamic mike with a built in pop-filter and is used by sports commentators. Since the mike is built into the headset, a constant mike-to-mouth distance—even when the announcer moves from place to place—is assured. The double earphones carry both the program audio and the director's cues.

Wireless Mikes

Wireless mikes (also known as RF mikes) are of great advantage in productions when talent must be free to roam and microphone cables would be a hindrance. Today, RF mikes are widely used in both studio and on-location productions. Camcorders even have built-in receivers for wireless mikes.

In a wireless microphone, the signal from the dynamic or condenser mike is converted to a low-power FM signal. The signal is transmitted in a more or less circular pattern through either an internal antenna (within the mike's case) or an external antenna (generally in the form of a short wire attached to a separate transmitting unit). Most RF mikes transmit on frequencies above the standard FM radio band in either the high VHF (very high frequency) range, or in part of the UHF (ultra-high frequency) band. Because UHF frequencies are also used by other radio services, professional wireless mikes allow selection of different frequencies. Frequencies between 450 and 960 MHz are now available, and are being used to full advantage by major manufacturers in desingning wireless mikes. A frequency can be chosen to eliminate the RF interference. Under optimum conditions wireless mikes can reliably transmit over more than a 1000 feet radius. If obstructions are present, especially metal objects, this distance is reduced to 250 feet or less.

The transmitter input gain is the single most important adjustment on any wireless mike system to insure optimum signal to noise ratio. The most difficult problem with properly adjusting the transmitter input gain involves duplicating the user's voice level in advance of the actual performance or use. Mike transmitters provide LED level metering for this purpose. Metering is also provided on the receivers, but often the transmitter metering is easier to use, since the receiver may not be accessible or visible from the transmitter during set-up.

Many of today's wireless microphones rely on analog companding to increase their dynamic range. Companding is a term used to describe the compression and re-expansion of an audio signal. (This was explained in Chapter 2.) The incoming audio signal is compressed in the transmitter, used to modulate the RF carrier, and again transmitted. Upon reaching the receiver, the audio is detected (separated from the RF signal) and expanded to its original amplitude.

The use of companding in wireless mikes allows greater dynamic range without the need for increased RF carrier deviation.

There are two types of wireless mikes: the self-contained (all-in-one) unit and the two-piece type. In the self-contained hand-held unit, the mike, transmitter, battery and antenna are all within the microphone. In a two-piece wireless unit, the mike is connected to a separate transmitting unit.

MICROPHONE CHARACTERISTICS

Frequency Response

The frequency response of a microphone determines the microphone's ability to pickup certain audio frequencies. The frequency response is typically measured over a range of 45 Hz to 18.000 Hz. The frequency response determines the quality of a microphone. Frequency response and the design of a microphone are determining factors for its use in production. For example, microphones having a "low frequency roll-off" control which reduces the microphone's response to low frequencies may not be well suited for production of Indian music with drum.

The human ear is less sensitive to both high and low frequencies, and as such these sounds must be louder to be heard. Accordingly microphones are desired to have a flat response over these frequencies Fig.3.14.

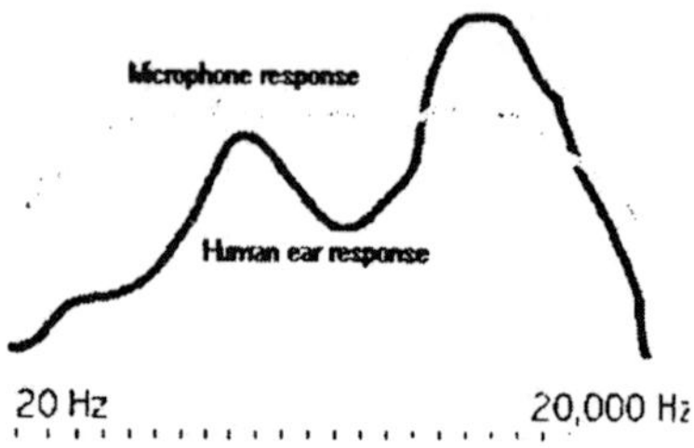

Fig. 3.14. Human ear response vis-à-vis Microphone Response Impedance.

Impedance

Another important characteristic of a microphone is its impedance, which refers to the electrical resistance in the circuit. This resistance is expressed quantitatively in ohms. Microphones of low impedance are preferred over high impedance mikes since long cables can be used with these mikes. Cheaper high impedance mikes can be deployed for short distance cables. For TV production, low impedance mikes are the de-facto standard.

STEREO, SURROUND SOUND AND QUADRAPHONIC

Stereo Sound

Human ability to locate a particular sound in three-dimensional world is based on interpreting the minute and complex time-difference between the perceived sounds from his left and right ears. If a sound comes from left side, it reaches the left ear a fraction of a second earlier than the right ear. There is also a slight difference in loudness between sound source located on the left as compared to the right. The interpretation of this subtle time difference (phase difference) and level difference helps in placing the sound in a three-dimensional perspective. In stereo production, two audio sources, one intended for left and one for right ear, are created. The slight differences between the tracks create the illusion of depth to the audio that is associated with stereo. The image below Fig. 3.15

illustrates a brief excerpt from the dual tracks of a stereo audio. They are similar, but not identical. Therefore, recording and playing back stereo signals requires two audio channels or tracks.

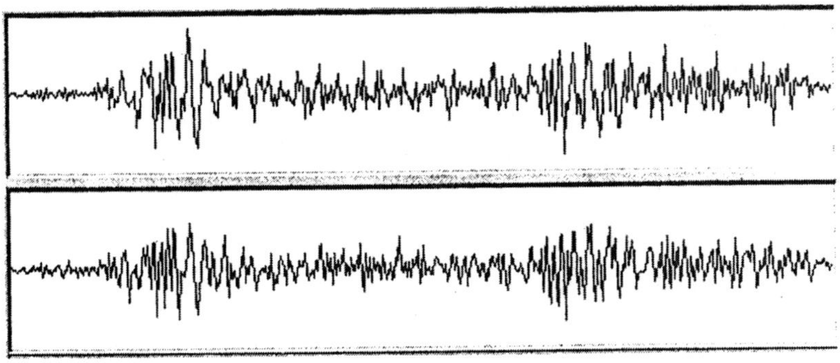

Fig. 3.15. Dual Track of Stereo Audio.

Stereo Mikes

There are two approaches for stereo production. One is the use of "all-in-one stereo mike", which is basically two mikes mounted in a single housing. The second method is to use two separate mikes. All-in-one stereo mikes are used for on-location recordings. This approach is limited in its ability to provide optimum stereo separation (a clear and distinct separation between the left and right stereo channels). For this reason, in more sophisticated production, many audio technicians prefer to use two separate mikes.

There are several methods used to create the stereo effect in production. In a synthesized stereo, the effect is simulated electronically. In this method, a monaural (one channel, non-stereo) sound is electronically processed to create a two-channel stereo signal by adding a slight bit of reverberation, or echo effect in one of the channel. True stereo sound is recorded with a stereo mike (with two sound-sensing elements) on two audio tracks. There are three basic techniques for placing mikes for stereo recording.

Spaced Technique

Two microphones are placed at equal distance, facing the sound source. However, the distance between the microphones is kept at three times the distance between the subject and the microphone for creating the required phase between the two channels.

X/Y Technique

Two microphones are placed, criss-crossed with their heads together but not touching each other. The angle, pointing away from the speaker, formed at the intersection of the two mikes, is kept between 60° and 120°.

M/S Micing Technique

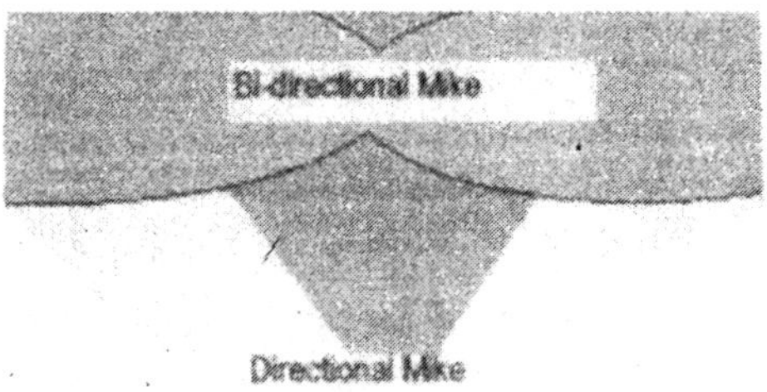

Fig. 3.16. M-S Micing Technique.

The most popular approach to recording stereo is the M-S micing technique. Here bi-directional and uni-directional (supercardioid) mikes are used together. The bi-directional mike's polar pattern, which is shaped somewhat like a figure-8, is aligned so that its areas of maximum sensitivity are parallel to the scene. This means that the areas of minimum sensitivity are oriented toward the centre of the scene and the camera placed at the bottom of the drawing in Fig. 3.16. The dead spot (area of reduced sensitivity) that is directed toward the camera ends up being an advantage because behind-the-camera noise is reduced. A directional mike is placed at the centre of the scene (where most dialogue originates).

The outputs of both mikes are fed through a complex audio

matrix circuit that uses the phasing differences of the mikes to produce the left and right channels. Adjustments to this circuit allow considerable latitude in varying the overall stereo effect.

Surround Sound

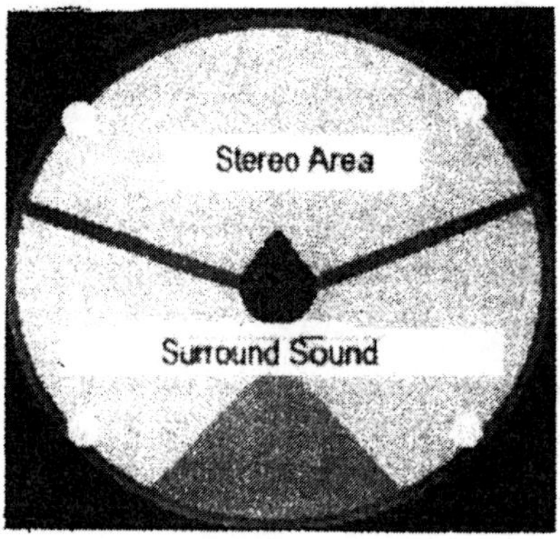

Fig. 3.17. Surround Sound.

Stereo covers an area of about a 120° in front of the listener. Surround sound and quadraphonic sound systems reproduce sounds in both the front and back of the listener—close to a 360° sound perspective. In playing back true surround or quadraphonic sound, at least four speakers are needed at positions indicated in Fig. 3.17.

Quadraphonic Mikes

Quad mikes have four mike elements within a single housing and are capable of detecting sounds in a 360-degree perspective. Typically, an upper capsule contains two mike elements and picks up sound from the left front and right rear. Another capsule, mounted below, picks-up sound from the right front and left rear. These are recorded onto four audio tracks. During post-production, the four audio tracks are fed through a computer, mixed with tracks of music and effects (M&E) to develop a full surround sound effect.

Foam Windscreens

Foam windscreens and pop-filters are intended to provide protection against low velocity moving air such as encountered from performers exhaling onto the mike. Foam windscreens also protect sensitive condenser mikes against the motion of room air caused by normal ventilation ducts as well as from the physical act of moving the microphone while mounted on a boompole. Many hand-held performance mikes come equipped with built-in wind or pop protection.

Foam windscreens also provide a quick means of altering the colour and appearance of microphones. Some sound mixers place a white band of tape around the front tip of the windscreen. This facilitates the camera operator identifying the microphone through even a dim viewfinder, where otherwise it might go unnoticed until brightly projected during dailies[10].

Television Cameras

The pick-up tube is the main element governing the technical quality of the picture obtained by the camera. Very large lenses characterized the first electronic cameras using iconoscope tubes, necessary to ensure enough light reaching the pick-up tube. The separation of the optical image from the electronic image gave the first improvement by allowing the target to be made smaller and enabling lenses to be used which had been designed originally for 16 and 35 mm motion-picture cameras. An early television camera used in 1937 is shown in Fig. 3.18. When colour television arrived, cameras again became bulkier owing to the need to accommodate the three tubes for the three primary colours. The tubes of the 1950s, of the image-orthicon type, were far from ideal for the purpose. However in Europe, the launch of colour services coincided with the arrival onto the market of the plumbicon tube and the few colour image-orthicon cameras in existence were rapidly put aside in favour of the new-camera. A modern television camera is shown in Fig. 3.19.

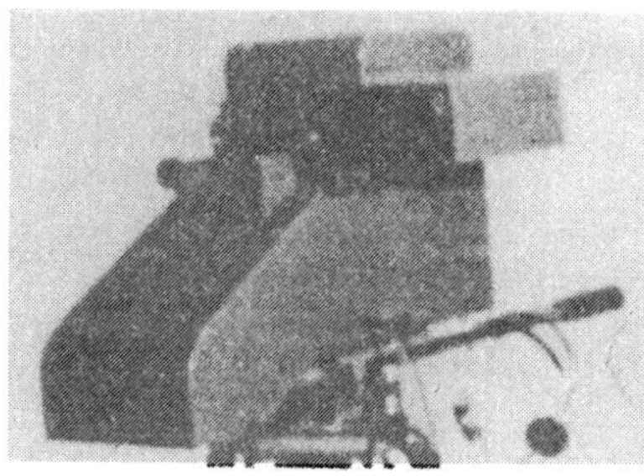

Fig. 3.18. An early television camera (1937).

Fig. 3.19. A modern television camera (1985).

Camera Tubes

The American scientist Viadimir Zworykin invented the iconoscope in 1929 Fig. 3.20. The tube had a vacuum-tight glass envelope. At its centre was a di-electric plate (mica), one surface of which was coated with a thin uniform layer of metal, the other with a mosaic of thousands of tiny metal elements which release electrons under the effect of illumination Fig. 3.21. The scene was focused onto this mosaic, which made up a series of mini-condensors. An electron beam bombarded these, one after the other. The amount of charge collected was proportional to the charge on each condensor, and therefore to the strength of illumination of each element. When it hit the target, the scanning beam ejected "secondary electrons" which formed a cloud around the point of impact and blurred the image.

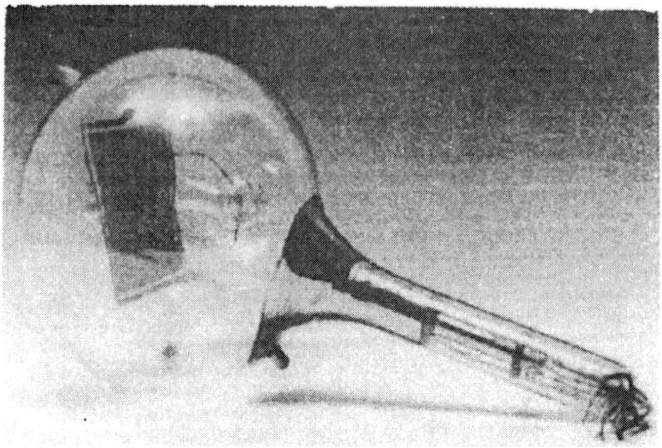

Fig. 3.20. Iconoscope glass envelop.

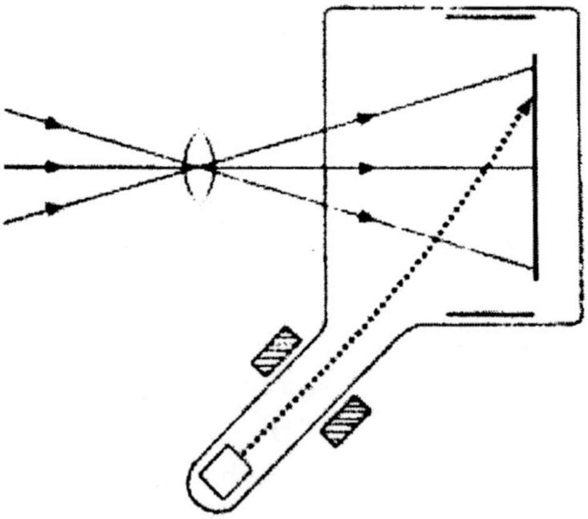

Fig. 3.21. Working of Iconoscope.

The *super iconoscope,* was invented in 1939. In this tube the optical image was separated from the electronic image. The optical image fell on a transparent photo-electric surface and produced a stream of electrons, which was focused on the mosaic scanned by the beam of electrons Fig. 3.22.

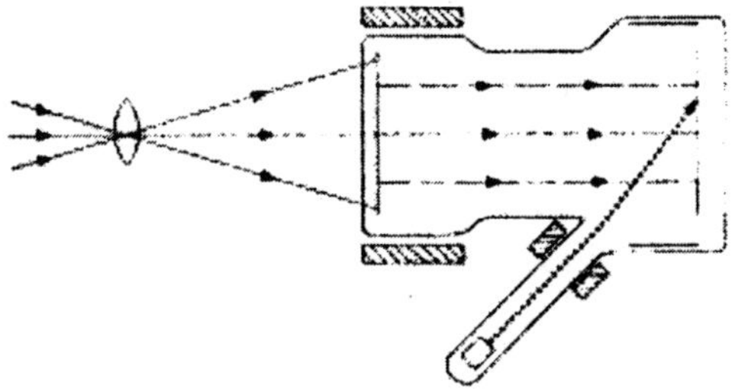

Fig. 3.22. Working of Super Iconoscope.

The orthicon, also invented in 1939, improved the scanning uniformity. The target was based on the same principle as the iconoscope (a mica sheet between a conducting surface and a photo-emissive surface) but the beam of electrons was aimed perpendicularly to the mosaic, which it hit at low speed (Fig. 3.23). The conducting surface was kept transparent to allow the beam to arrive at the appropriate side of the mica sheet.

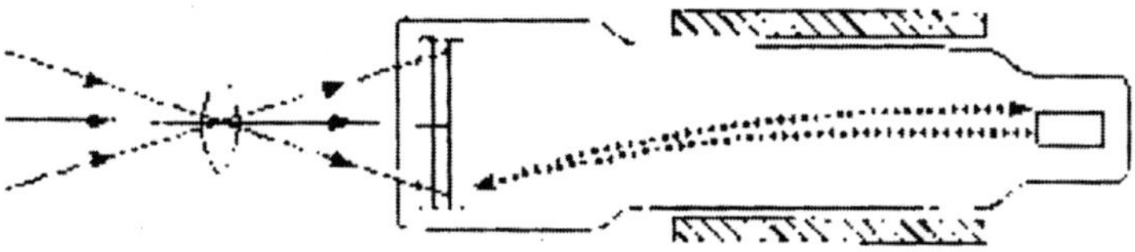

Fig. 3.23. Working of Orthicon.

The *image-orthicon* Fig. 3.24 was developed in 1946 by RCA. The front of the image orthicon contains a screen called a photo-cathode that releases electrons when light from the camera lens strikes it. Bright parts of the scene knock out more electrons than dim parts do. Another screen behind the photo-cathode, called the target, attracts the released electrons, and a positively charged electronic image of the scene forms on the target. The image consists of highly and weakly charged spots that correspond to the bright and dim areas of the scene. A beam of electrons then scans the target, which absorbs electrons from the beam in proportion to those knocked out by the image. The remaining electrons in the beam bounce from the target. At the rear of the image orthicon, a device called an electron multiplier strengthens the returning beam several thousand times. This beam becomes the television picture signal. The development of the image orthicon in the early 1940s, followed the iconoscope a device that accomplished a similar purpose, but required an extensive lighting system to operate properly.

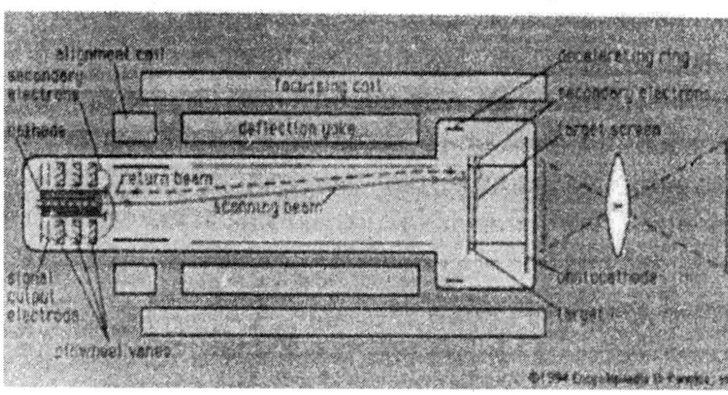

Fig. 3.24. Image-Orthicon.

Source(s): *Image: Encylopedia Britannica, Text: jackb@pobox.com*

The *Vidicon* was developed in 1951. It uses a photo-conductive target in place of a photo-emissive one. The photo-conductive material whose electrical resistance varies according to the strength of illumination, is placed in direct contact with the metallic coating where the signal is produced. This layer is maintained at a positive voltage compared with the cathode but the photo-conductive material acts as a valve which allows, or prevents, the electrons to pass through, depending on the amount of light Fig. 3.25.

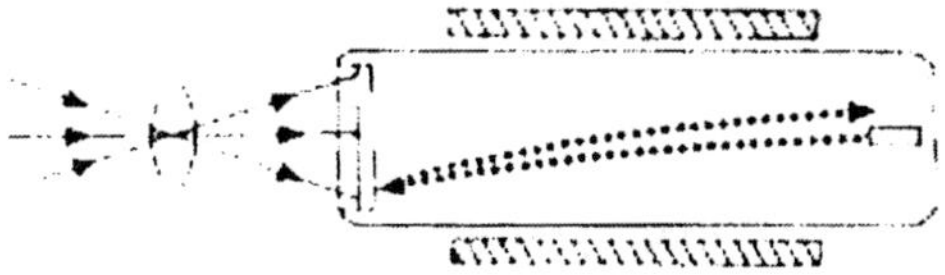

Fig. 3.25. Vidicon.

All modern tubes are variants of the Vidicon. Philips launched the Plumbicon in 1962. The Plumbicon target contains lead oxide. Hitachi produced the Saticon in 1973 with target of selenium, arsenic and tellurium.

New Imaging Devices

Within the TV camera, the imaging device is the most important component. There are two types of imaging devices, the tube and CCD (charged-coupled-device). As discussed in the preceding para, the picture tube consists of a round cylinder with a light conductive face on one end. When light is shined on the light-sensitive side, an electronic beam scans the image that is formed on the other end of the tube and converts the image to an electronic signal. Tube cameras albeit with superior resolution, are fragile, need regular maintenance and are not suited for field production. Today, because of CCD technology, tube cameras are hardly used.

CCD—are either assembled as a linear or two-dimensional array of light sensitive elements. The CCD is actually an integrated circuit, a computer chip, with light sensitive pixels. Light is converted to an electrical charge proportional to the light impinging on each cell. The cells are coupled to a scanning system, which, after analogue to digital conversion, presents the image as a series of binary digits.

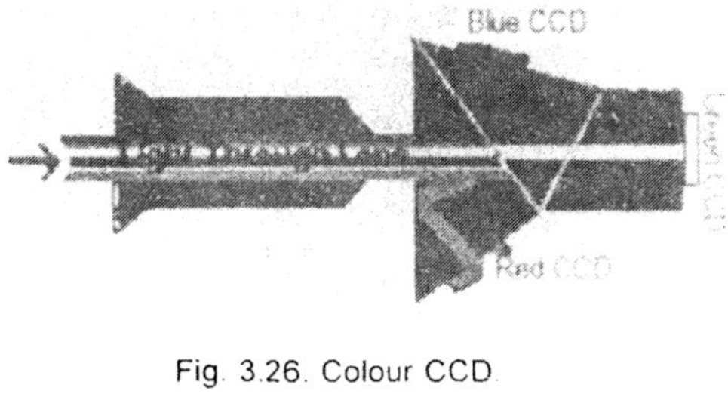

Fig. 3.26. Colour CCD

Although consumer cameras may have only one CCD, professional cameras incorporate 3 chips (3 CCD) to gather the image. The full-colour image "seen" by the colour TV camera lens goes through a beam-splitter that separates the full-colour picture into its red, blue and green components Fig. 3.26. Thus, each CCD picks up the three colours that make up a TV image (Red, Green, and Blue).

Early CCD arrays were unable to reproduce a wide range of brightness but they now offer low noise high-resolution imaging. Beyond current broadcast television resolution, defect-free two-dimensional CCDs are now produced for HDTV.

CAMERA CHARACTERISTICS

One characteristic of a TV camera is its signal-to-noise ratio (S/N ratio). Typically, a pro (professional) camera would have a S/N ratio of > 60 dB. CCD cameras tend to have a higher S/N ratio than tube. Therefore, CCDs are preferred for field applications since they can be operated at a variety of light levels. The contrast range of the camera refers to the differences in brightness within a TV image. This range should not exceed 30:1. In other words, the brightest portion of an image should not be 30 times greater than the darkest area. If this ratio is exceeded, then the detail in the image will be sacrificed. Resolution (which is actually the vertical resolution or the number of dots on each horizontal line of the picture), also affects the final image. Most consumer camcorders have a vertical resolution of 300 to 350 vertical lines, while professional ENG and studio cameras can produce images of up to 850 vertical lines. A comparison of CCD and tube technologies is given in Table 3.1.

Table 3.1. Comparison of CCD and Tube Technologies

Feature	*CCD*	*Pickup Tube*
Camera Size/Weight	Compact and lightweight	
Ruggedness	Solid state, resistant to shock and vibration	Glass envelope is fragile
Power	30% less power than tube	
Start-up	Immediately ready for shooting	Needs warm-up time
Registration	Factory Pre-set	Auto/Manual
S/N ratio	< 62 dB	< 57 dB
Resolution	Excellent	Superior
Lag/Comet-Tailing	None	Occurs with reflective surfaces
Burn-in	Not a problem	Can occur when pointed at a light
Smear	Vertical banding around bright objects	None
Aliasing	Jaggedness on curves	None
Maintenance	CCD never needs replacement	Tube needs regular replacement and care

Source : Television Production by Alan Wurtzel and John Rosenbaum.

CAMERA DEVELOPMENT

A modern XL1 Pictures Processed w/ Bvp4 Plus camera is shown in Fig. 3.27.

Fig. 3.27. A modern CCD Camera.

CAMERA LENSES

A camera lens collects the light from the subject or scene. A camera lens may in fact hold over 16 different glass elements, necessary for image reproduction. Sophisticated zoom lenses reduce the amount of light reaching the CCD or tube by 20 to 25 per cent. Hence, zoom lenses require higher light levels to operate.

Iris (Aperture)

The iris or aperture is a series of metal leaves, which open and close to vary the amount of light entering the camera. The size of the iris opening is known as an F-stop. F-stop numbers range from f/1.4 to f/22. As the f-stop number increases the amount of light entering the camera decreases by 1/2. Lenses are rated at the lowest f-stop number. Generally when using a camera in the field, a lens with a wide f-stop is desired.

Focal Length

The focal length of a lens refers to the distance from the "optical centre" of the lens to the point where the light rays converge at the rear of the lens. This distance is measured in millimeters. This point where the rays converge is ideally the CCD or tube. "Short" lenses or "long" lenses refer to focal length. As the focal length decreases, the image becomes wider and as the focal length increases, the image becomes narrower or magnified. Zoom lenses provide variable focal lengths. Optical devices such as range extenders "extend" the focal length.

Zoom Lens

The zoom lenses are of two types, the manual zoom and the servozoom. In a manual zoom lens, a crank located on the right pan-handle controls the focal length. To operate a servozoom, a thumb switch located on one of the pan bars on the tripod is activated.

Filters

Filters are used to adjust the colour of the light entering the camera. Most broadcast cameras are equipped with filter wheels. A neutral density filter is basically a gray filter that does not affect the colour quality of the light. It is used on ENG/EFP cameras where light conditions are too excessive for the camera (i.e. bright sunny days or snow scenes). Colour correction filters are used with pro cameras to correct the daylight or fluorescent lights. Besides the colour correcting filter, a diffusion filter is used for creating a blurry, dreamlike effect (like videotaping of weddings). These are also used to soften the edges of the image while keeping the middle sharp.

Camera Viewfinder

The camera viewfinder shows the image being picked up by the lens of the camera and some other critical information. The viewfinder either uses a CRT (cathode ray tube) or a flat LCD (liquid quartz display). The viewfinders of studio cameras typically use at least 7 inch displays, however, for camcorders these are much smaller with a magnifying eyepiece. LCD viewfinders are helpful in shooting at very low or high angles. The main disadvantage of the LCD display is that the image loses contrast and brightness in bright light, and the camera focusing becomes difficult. CRT viewfinders suffer from electrical focus problem and the occasional lack of proper image centreing. EFP camera manufacturers have added an array of status indicators to viewfinders. Viewfinder status indicators may include the following items:

- a tally light (indicating that tape is rolling or that camera is "on the air")
- low battery warning
- minutes of tape remaining
- colour balancing may be needed
- low light, insufficient exposure
- low-light boost (gain selector switch) circuit in operation
- indoor/outdoor filter in place
- zoom lens setting indicating how much further you can zoom in or out
- auto/manual iris status
- audio level meter
- tape footage counter
- "zebra pattern" for monitoring and setting video levels
- super imposed masks for safe area, and 4:3 and 16:9 aspect ratios.
- the presence of customized camera set-up profiles (to
- accommodate specific types of subject matter)
- camera warm-up diagnostics

CAMERA MOUNTING EQUIPMENT

A tripod with a mounting head is used to hold the camera. There are two types of heads designed for ENG/EFP. Friction heads use a spring to control camera movement. They are not meant for use with large studio cameras. Fluid-head is filled with a viscous fluid that controls the movement. This head provides very smooth camera movements. Three types of moment—pan, tilt, and drag are possible.

Camera Pedestals

The camera pedestal is used to change the height of the camera. In the counterweight pedestal, the pedestal is counterbalanced by weights. This allows the moment of camera upwards and downwards by a slight pressure from the operator on the centre-lifting ring. The pedestal mounted on wheels has two motions, crabbing (steering in parallel) and tracking in which the first wheel steers, the rear wheels remain in parallel. Pneumatic pedestal using compressed air are also used.

OTHER MOUNTING EQUIPMENT

Crab dolly: The camera is mounted on a large mechanical boom. The operator sits on a seat. Another person steers the dolly across the floor. This provides smooth dollies across the studio.

Camera-crane: The camera is mounted on an arm extending outward. The arm is connected to 4 pairs of wheels. The arm is hydraulically and remotely controlled. The camera can be lifted to over 20 feet from the floor.

Jibs: These are mini-cranes. The camera is mounted at the end of the jib and the operator can control pan, tilt, zoom, and jib operations using servocontrols.

STABILIZERS AND BODY MOUNTS

Body stabilizers such as the Steadicam allow for rock steady shots. Using the Steadicam the operator can create smooth dollies and great shots. The Steadicam is mounted on the person with the camera attached to the device.

AUDIO RECORDING AND PLAYBACK DEVICES

Turntables and Reel-to-Reel Tape Machines

Turntables and Tonearms

A turntable is used to play a record. A drive system rotates the disk at the desired speed. The history of evolution of record drive mechanisms starts from the days of hand-cranked cylinder machines, through the age of spring-wound phonographs with mechanical governor for speed control to the age of electrically driven machines having the sophistication of modern computers. A tonearm with a cartridge stylus is used to track the groove on the record and move it across the record until the end of the groove is reached and the shutoff mechanism is triggered or a record change takes place. The records are subject to scratches and noise and as such the turntables are slowly being replaced by CDs.

HALF-TRACK REEL

The workhorse of the radio industry, these are used widely in the Radio stations. Radio spots are still distributed on 7 1/2 ips 1/4" reel without noise reduction. With tape formulations ranging from ferric (iron rust) to chromium dioxide to pure un-oxidized metal particles, tapes have come a long way. A good Studer, Ampex, or Tascam recorder with external noise reduction is a very clean and good sounding medium. Many recordings are still mastered this way, but the half-track reel is gradually being replaced by the DAT recorder.

MULTI-TRACK REEL TO REEL

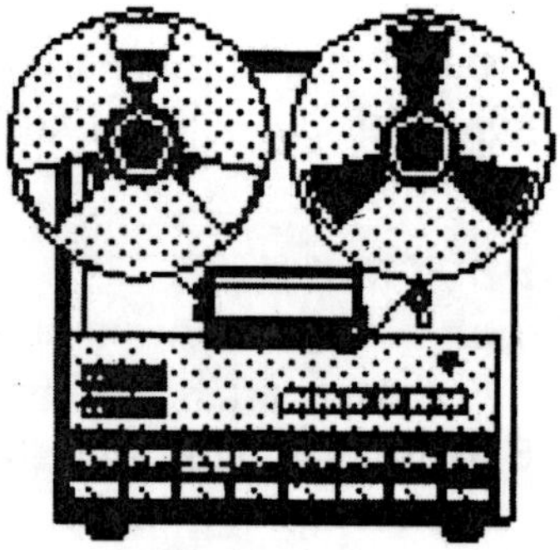

Fig. 3.28. Multi-track reel-to-reel.

These are the tape machines having 4 to 24 channels using 1/4 inch to 2 inch tape. Noise reduction is usually through internal Dolby C. The recording is of 30 minutes duration on a 2500 ft tape at 15 inches per second. Fig. 3.28.

Multi-Track Cassette

Multi-track cassette is widely used in the studios. Tracks range from 4 to 8 on a standard cassette, and basic mixing board functions are usually included. Internal noise reduction is built-in and used to offset the noise of the cassette medium.

Standard Cassette

Standard cassette tape recorders are now available with noise reduction such as Dolby B, Dolby C, dbx type II, Dolby A, dbx type I, and Dolby SR, plus new ones like Dolby HX-Pro and Dolby SL.

New Machines

Although records and reel-to-reel analog tape machines were used for several decades for audio production, they are gradually being replaced by audio carts (cartridges), CDs (compact discs) and DAT (digital audiotape) machines.

"Vinyl," a term that refers mostly to LP records has a number of disadvantages over the newer recording media--primarily the propensity to get scratched and worn. Both of these maladies lead to surface noise.

Unlike records or vinyl, some of the newer media can be electronically cued, synchronized and precisely controlled. These are all important in precise post-production audio work.

CART MACHINES

Cart machines (cartridge machines) contain a continuous loop of 1/4 inch audiotape within a plastic cartridge. Thus in a cart machine, the tape moves in a continuous loop. This means that unlike an audiocassette which has to be rewound after a wait, in a cart, the beginning point recycles again, when the tape is automatically stopped. The continuous loop feature differentiates carts from cassettes both in audio and videotape recording.

The basic controls on cart machines are start, stop, play and record.

Inaudible tones are recorded on the tape. These are used by the machines to mark start and stop points. To speed cueing, some machines accelerate tape transport between the point when one segments ends and the start of the next segment.

Most carts record and playback 30 and 60 second segments (used for commercials and public service announcements) or about three minutes. Carts that record up to one hour are available.

COMPACT DISCS (ROM)

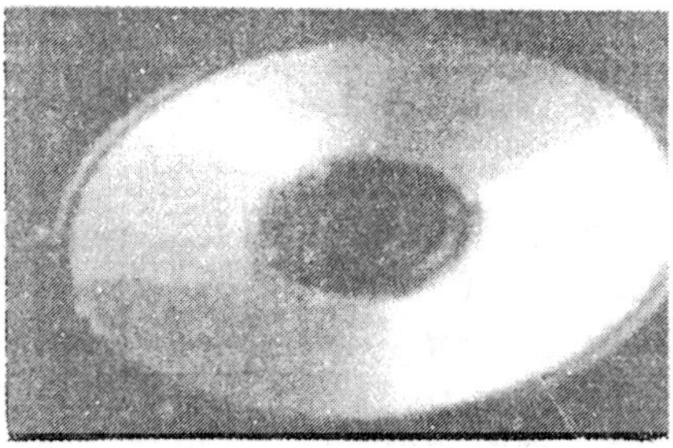

Fig. 3.29. Compact Disc.

Introduced in 1983, CDs (compact discs) are now the preferred medium for pre-recorded music and sound effects due to their superior audio quality, ease in control and small size. CD-ROM (Fig.3.29) is cheap, durable, unaffected by magnetic disturbance and free of most hazards that prey storage media. Ultimate in audio excellence, the CD format is a real-time, 16 bit linear pulse code modulation (PCM) process and supports a wide bandwidth of 20 kHz. CD are 12 cm in diameter, just over a mm thick, weighs about 16 grams and store upto 680 MB of data consisting of text, images, sound and graphic. The data is stored in the form of microscopic pits about half a micron wide, arranged in a single spiral track starting from inside to outside (total length of about five kilometers). A CD contains about 5 billion pits. When a CD is played, a laser beam is used to illuminate the microscopic digital pattern encoded on the surface. The reflected light which, is modified by the digital pattern, is read by a photo-electric cell. Its principal limitation is read only nature and the enormous amounts of storage space (Physical) and circuit bandwidth it occupies.

The width of the track is 1/60th the size of the grove in a LP record, or 1/50th the size of a human hair. If "unwound" this track would come out to be about 3.5 miles (5.7 km) long. Although the overall diameter of a typical audio CD is only about five inches (12.7 centimeters) across, CD's are able to hold more information than a 12 inch, LP (long playing) phonograph record can on both sides. And, the frequency response (the sound's pitch from high to low) and dynamic range (the audio range from loud to soft that can be reproduced) are significantly better. The technical specifications of CD are given in Table 3.2.

Table 3.2. Technical Data for CD system

Disc	
Diameter	120 mm
Thickness	1.2 mm
Reading Speed	1.2 to 1.4 m/s
playing time	max. 60 min (stereo)
Signal	
sampling frequency	44.1 KHz
Quantization	16 bit linear per L/R channel
Data rate from disc	4.3218 Mbit/s
Audio performance	
Frequency range	20 Hz to 20 KHz; +- 3 dB
S/N	>90 dB

CD playbacks can be programmed for instant starts at precise points. This is a significant advantage in audio production when an audio operator or a video edit controller must start musical transitions and audio effects at pro-programmed points.

CD-R (CD RECORDABLE)

Although the discs containing permanently recorded audio are most common, there are also RCD's (recordable compact discs). With these, it is possible to repeatedly record and playback material on the same disc. Compact Disc Recordable (CD-R) or CD-WO (Compact Disc Write Once) available since 1989, use a layer of organic dye and a very thin layer of reflective gold on a pre-grooved polycarbonate disc. The dye polymer, when exposed to a cutting laser beam, reacts with the polycarbonate to form a pit. This pit is read as any other CD-ROM. The CD-R can be written in multiple session. Discs with storage capacity of 550 MB are now available.

COMPACT DISC INTERACTIVE (CD-I)

CD-I is capable of playing upto 72 minutes of moving pictures on full screen with audio of CD quality. A typical full motion CD-I disc allocates 1.2 Mbit/sec for video and 0.2 Mbit/sec for audio in stereo quality. In order to compress the audio/video to this bit rate (1.4 Mbit/sec), the MPEG coding is used. During playback, the audio/video is decoded. The main features are 50/60 Hz compatibility and high quality full motion play back of 72 minutes duration.

CD Defects and Problems

If the surface of the CD is sufficiently warped due to a manufacturing problem or improper handling or storage, the automatic focusing device in the CD player will not be able to adjust to the variation. The result can be mistracking and loss of audio information.

Automatic Error Correction

Manufacturing problems, and dust and dirt on the CD surface causes a loss of digital data. CD players attempt to compensate for the loss of a signal in three ways: error-correction, error concealment (interpolation) and muting. Error-correcting circuitry within the CD player can detect momentary loses in data (dropouts) and, based on the existing data at the moment, supply the missing data. If the loss of data is more significant, error-correcting circuits instantly substitute data that blends in with the existing audio. If this type of error concealment has to be invoked repeatedly within a short time- span, a series of clicks or a ripping sound is noticed. Finally, if a large block of data is missing or corrupted, the CD player simply mutes (silence) the audio until good data again appears.

DIGITAL VIDEO DISC (DVD)

DVD are capable of storing about 2 to 10 GB data on a single disc. Decreasing the size of the pits as well as the distance between them and utilizing advanced compression techniques allows this. A single DVD of 12 cm can store two to five hours of high quality video.

It was desirable to have two or more hours of recording (one movie) on a 3 Gbit DVD. With the requirement of 6 Mbit/s of MPEG-2 in constant bit rate (CBR), about one hour of recording is only possible. As a solution, a variable bit rate (VBR) coding which permits 3 Mbit/s with sufficient picture quality is used. Coding and multiplexing is still based on MPEG-2 method (ISO/IEC-13818-2) which supports VBR . Main-Profile at Main-Level for Video and Program Stream for Multiplexing is used. Audio signal is coded in CBR.

DAT, RDAT and ADAT Types I and II

Developed in 1985, digital audiotape had two recording methods viz. R-DAT (rotary head digital audio tape recording) and S-DAT (stationary head digital audio tape recording). Finally R-DAT was standardized. R-DAT uses digital format with a sampling frequency of 48 KHz and linear 16 bit quantization.

With an absolute tape speed of 8.15 mm/s, the R-DAT requires only 60 meter of tape for two hour playing time. The digital audio signals yield a data stream of about 1.5 Mbits/sec. Special sections within the track are reserved for auxiliary information viz. programme titles and time-codes. The main disadvantage of DAT is lack of random access capability.

DATs are capable of audio quality that exceeds the quality of CDs. DAT has been finding wide acceptance for motion picture as well as television production. DAT systems use a head-wheel that spins at 2,000 rpm, similar to what is found in a video cassette recorder. Various types of data can be recorded with the audio. Examples are time code and the MIDI machine control data used in sophisticated post-production audio work. The 2 inch by 2-7/8 inch DAT cassette contains audiotape 3.81mm wide and is about two-thirds the size of a standard analog audiocassette. The two-hour capacity of a DAT cassette is 66 per cent greater than a standard 80 minute CD.

DAT exists in several forms. The one for general consumer (home) use is just referred to as DAT. RDAT (recordable digital audio tape) is for professional applications, as are the very high quality ADAT machines (types I and II).

DAT Time Code

The DAT time-code system, referred to as the IEC Sub-code Format, is similar to the SMPTE time-code. It ensures that tapes recorded on one DAT machine can be played back without problems on any other machine.

Video Tape Recorder

Videotape recording process was invented in 1956. The system had four heads on a disc rotating perpendicularly across the width of the tape, thus tracing an oblique track pattern. The tape was 50.8 mm (2 inches) wide. It took four passes of video heads to scan a complete video picture recorded on a two-inch wide videotape. Referred as the 2 inch quad system, it was used for several decades. Recording on VTR is done through video heads across which the tape is moved resulting in magnetic traces on the tape's coating corresponding to the video signal. The heads, themselves, spin at a high speed to record the very high frequencies associated with video. Helical-scanning 0 is now used in all video recorders: each track contains one entire field (or a major part of it), and the tape can be "read" at different speeds, even when it is stationary. The magnetic tape is 25.4 mm (1 inch) wide.

DVTR—Digital Video Tape Recorder

First DVTR (Digital VTR) for commercial use was available in 1986, working to the CCIR 601 component digital standard and the associated D1 standard for DVTRs. It uses 19 mm cassettes recording 34, 78 or (using thinner tape) 94 minutes. Today many DVTR formats are available. D2 and D3, both recording composite signals, are designed mainly to replace C format analogue machines; DCT and Digital Betacam both make use of mild data compression (around 2:1) to record the CCIR 601 video. D5, like D1, records the full, uncompressed CCIR 601 signal but on 1/2 inch tape cassettes. At least one more format is on the horizon—DVC will record compressed CCIR 601 onto 6.35 mm tape. The format details shall be described further in the series.

Multiple generations on DVTRs do not suffer from degradation due to tape noise etc. However the tape is subject to wear and tear. The possibility of this producing errors and drop-outs necessitates error concealment circuitry. In extreme cases multiple passes can introduce cumulative texturing or other artifacts.

Video Recording Formats

Over the years, many video-recording formats have been introduced for television production. A comprehensible table of facts regarding the different formats is given in Table 3.3.

Table 3.3. Video Recording Formats

Format	*Tape Width*	*Tape Type*	*Signal Type*	*Luminance S/N Ratio*	*Chrome S/N Radio*	*Horizontal Resolution*	*Total Number of Audio Tracks*
VHS	1/2 in.	Oxide	Composite	46 dB	45 dB	240 lines	1 linear; 1 stereo AFM (Hi-Fi)
8mm	8mm	Metal Pasticle	Composite	46 dB	45 dB	260 lines	1stereo AFM; 1 stereo PCM
S-VHS	1/2 in.	Oxide	Y/C	47 dB	46 dB	400 lines	1 linear; 1 AFM (Hi-Fi)
Hi-8	8 mm	Metal Particle	Y/C	47 dB	46 dB	400 lines	1 stereo AFM; 1 stereo PCM

3/4 in. SP	3/4 in.	Metal Particle	Composite	47 dB	48 dB	340 lines	2linear
Betacam	1/2 in.	Oxide	Component Analog	48 dB	50 dB	300lines	2linear
Betacam SP	1/2 in.	Metal Particle	Component Analog	51 dB	53 dB	344 lines	2linear; 2 AFM
M-II	1/2 in.	Metal Evapo rated	Component Analog	52 dB	50 dB	344 lines	2 linear, 2 AFM
1-inch Type C	1 in.	Oxide	Composite	52 dB	52 dB	360 lines	3 linear
D-1	3/4 in.	Metal Particle	Component Digital	N/A	N/A	460 lines	4 digital, 48KHz
D-2	3/4 in.	Metal Particle	Composite Digital	N/A	N/A	450 lines	(4) 20 bit
D-3	1/2 in.	Metal Particle	Composite Digital	N/A	N/A	450 lines	4 digital, 48 KHz
D-5	1/2 in.	Metal Particle	Compo-l nent Digita	N/A	N/A	N/A	4 digital, 48 KHz
Digital Betacam	1/2 in.	Metal Particle	Component Digital	N/A	N/A	< 500lines	(4) 20 bit
DV or DVC	6 mm	Metal Evaporated	Component Digital	N/A	N/A	< 500 lines	(2) 16 bit (4) 12 bit PCM
DVCPO (D-7)	6 mm	Metal Particle	Component Digital	N/A	N/A	< 500 lines	(2) 16 bit; (4) 12 bit PCM
DVCAM	6 mm	Metal Evaporated	Component Digital	N/A	N/A	< 500 lines	(2) 16 bit; (4) 12 bit PCM
Digital S	1/2 in.	Metal Particle	Component Digital	N/A	N/A	< 500 lines	(4) 16 bit; 48 kHz PCM
Betacam SX	1/2 in.	Metal Particle	Component Digital	N/A	N/A	< 500 lines	(4) 20 bit

VHS

VHS was originally designed for home use where it is still used. Standard VHS is characterized by low resolution extremely limited dubbing capabilities, and poor colour reproduction. Therefore, it is commonly used in non-broadcast facilities and for off-line screening.

8 mm

Kodak introduced 8 mm format in 1980s. The 8 mm format's resolution is a little better than VHS. It uses a high quality metal tape making it suitable for other formats. 8 mm is not used for TV production because of its low quality.

Hi-8

Hi-8 format was introduced by Sony in 1989 as an acquisition format, which was later adopted into editing. Sony expected users to dub footage recorded on Hi-8 to Betacam SP or 1 inch for editing.

S-VHS

S-VHS format, evolved from VHS, has a much higher resolution compared to VHS. Therefore, S-VHS can survive multiple dubbings without being degraded. S-VHS is an appealing format to the "prosumer" market because S-VHS machines can play back VHS tapes. However, S-VHS (as well as Hi-8's) S/N ratio is only slightly better than their "lower-end" counterparts.

3/4 inch and 3/4 inch SP

The 3/4 inch format was invented in the early 1970s and was very popular throughout that decade as well as most of the 1980s. Betacam SP and M-II have replaced it as broadcast formats. The major disadvantage of the format is its quality and the larger size of the tape as compared to1/2 inch and 8 mm formats. However, it is still being used in cable facilities and post-production houses. The SP format is an enhancement to the original 3/4 inch. SP has better resolution as well as a better S/N ratio.

Betacam and Betacam SP

Betacam SP, introduced by Sony, provides outstanding colour reproduction as well as resolution. This is the most popular video format used in the professional and broadcast industry. The video is processed in component form meaning that the colour and brightness informations are recorded on separate video tracks. This makes possible additional generations for layering/effects. Compared to the original Betacam format (introduced early 1980s), Betacam SP uses higher quality metal tape and delivers Hi-Fi sound.

M-II

Panasonic designed the M-II format for NHK, Japan's major television network. Compared to Betacam SP, M-II has a slower recording speed but uses better tape (metal evaporated compared to metal particle).

1 Inch Type C

Fig. 3.30. 1 inch format VTR.

1 inch Type C was one of the first video formats introduced in the industry (Fig. 3.30). It has been used throughout the 70s, 80s, and 90s. The 1 inch format has outstanding slow-motion and freeze-frame capabilities since a complete video field is written on every head scan of the tape. Due to the introduction of better component analog and digital formats, it is no longer used for day to day broadcasts. Instead, 1 inch is mostly used for broadcasting movies and in sporting events for playing back slow-motion instant replays.

D-1

D-1 is a component digital format introduced by Sony in 1986. The width of the tape is 3/4 of an inch, with the resolution of about 460 lines. It is considered as a quality reference because of its supreme quality. However, D-1 recorders are extremely expensive and impractical for industry use.

D-2

D-2, a composite-digital format, was developed by Ampex around the same time that D-1 was introduced by Sony. The width of the tape is 3/4 of an inch and the resolution is about 450 lines. Other formats because of tape cost, size, and impracticality have superseded the D-2 format.

D-3

Panasonic introduced D-3 in 1991. It has been considered as one of the first successful digital formats in the industry. The tape width is 1/2 inch making it possible to build D-3 camcorders.

D-5

Panasonic introduced D-5, a component digital format, in 1993-1994. D-5 could also play back D-3 tapes. The overall picture quality is better than the older D-3. Currently, NBC, NHK (Japan), and PBS are the big networks using D-5.

Digital Betacam

Digital Betacam, a component digital format using 10 bit 4:2:2 sampling, was introduced by Sony in 1993 as a successor to Betacam SP. The format has been popular in film transfer because of its excellent quality and its ability to record video with a 16:9 aspect ratio.

DV or DVC

Introduced in 1995, DV or DVC is the first major high quality video format for the consumer market. The format uses a 5:1 compression, M-JPEG algorithm. Some popular camcorders that utilize the DV format include the Sony VX-1000 and Canon XL-1.

DVCPRO

Panasonic introduced the DVCPRO format simultaneously when the regular DV format was introduced. It is much more affordable and possesses a quality, meeting or exceeding Betacam SP. Fig. 3.31.

Fig. 3.31. DVCPRO AJ-LT75 Laptop.

DVCAM

DVCAM was introduced by Sony as their professional DV format. The DVCAM recording format incorporates a higher tape speed compared to regular DV, but it is slower than DVCPRO. To compensate for the slower tape speed, DVCAM uses metal evaporated tape.

Digital S

Digital S was a format created by JVC. Compared to DV, DVCPRO, and DVCAM, Digital S has two advantages: (1) it uses 4:2:2 sampling to record digital video (like D-1), (2) Digital S VTRs can playback S-VHS tapes. The Digital S format is more robust than DVC, DVCPRO, and DVCAM. Technically, Digital S is better than the DV formats, which only use 4:1:1 sampling. As a result, DV does not produce sharp chroma keys. However 4:2:2 allows better colour sampling and hence better keys. The format uses half tapes looking similar to S-VHS tapes. In addition, Digital S has pre-read capabilities (the ability to record and playback at the same point on the tape track—useful for A/B rolling with only two decks) as a high-end Beta SP deck.

Betacam SX

Betacam SX was developed by Sony and introduced in 1996. Betacam SX uses 4:4:2 MPEG 2 sampling and 10:1 compression making the image quality close to digital betacam. Betacam SX allows the videomaker to playback and record on analog Betacam SP cassettes.

Video Servers

The video server is the storage device which includes computer network and SCSI (Small Computer System Interface) to distribute and archive compressed video as data, avoiding generation losses and taking advantage of standard computer network and tape transports.

The unprecedented expandability of the video server is made possible through the invention of a low cost scaleable interconnect called the Data Transport Commutator which makes all of the arrays in a system appear as a single pool of video and audio to each channel. Only a single copy of any video or audio segment is needed and all channels have simultaneous access to it. This interconnect is a high speed switching fabric which allows the server to scale to dimensions never before possible in a server architecture.

A video server can be represented by five key components: (i) RAID disk array, (ii) a real time controller for traffic management on the bus, (iii) I/O cards, (iv) high speed internal data bus, and (v) disk scheduling that configure the video server for specific applications.

Video server must contain storage, and a common choice today is magnetic hard disk. While in nonlinear editing, a single disk may be used to store the compressed video as it is digitized. Broadcast applications differ in two fundamental respects. One, broadcast applications demand multiple simultaneous channel accessing of the same material and two, data protection. Disk arrays with RAID satisfy both these requirements.

Mixing of Audio

Audio sources can be of two categories, based on their level.

- mike-level inputs
- line-level inputs

Mike-level inputs handle the extremely low voltages associated with microphones while line-level inputs are associated with the outputs of amplified sources of audio, such as CD and tape players.

During a production, the various sources of audio must be carefully controlled and blended. If audio levels are too high, a level distortion can result; and if levels are too low, noise can be introduced. Besides, there are creative needs to carefully control and blend audio levels for the optimum effect. Maintaining of optimum levels throughout a production is commonly referred to as *riding gain.* The control of audio signals is normally done in a Radio / TV studio or production facility, with an audio board or audio console or audio mixer. All mixing systems have pretty similar functions. They amplify and equalize signals that are then routed to various feeds and speaker systems.

AUDIO MIXER CONSOLE

Audio consoles perform following functions.

- amplify incoming signals
- allow level (volume) adjustments for individual audio sources, with the help of VU meters and provides means for changing the character of an individual sound—making it louder, softer, brighter, fatter, etc.
- provide for the aural monitoring of individual sources, as well as the total audio mix
- allow smooth mixing (blending together) of multiple audio signals either recorded to tape or for a live audience.
- route the combined effect to a transmission or recording device

Inside an audio system, all audio sources are changed to line-level and controlled in the same way. In addition, sophisticated audio control systems also allow left-to-right "placement" of stereo sources, shaping frequency characteristics of sounds, adding reverberation to audio, etc.

Audio mixers and consoles use either selector switches or faders to control audio. Selector switches simply allow selection from a variety of audio sources and switch into an audio channel. Faders (volume controls) also referred to as attenuaters or gain controls are either linear or rotary in design. Linear faders are also known as vertical faders or slide faders. Rotary faders are commonly called pots (for potentiometers). A simple fader system is shown in Fig. 3.32. The input selector switches at the top of each fader can switch between sources viz. Microphones, CDs, videocassette machines, DATs, satellite feeds, etc. The selector switch at the bottom of each fader, switches the output of the fader between cue, audition and program. Cue is a term used for finding the appropriate point in a musical selection on an audio source. Audition allows an audio source to pass through an auxiliary VU meter to full range speakers so levels can be set and audio quality evaluated. Program sends the audio out to the master gain control to be recorded or broadcast.

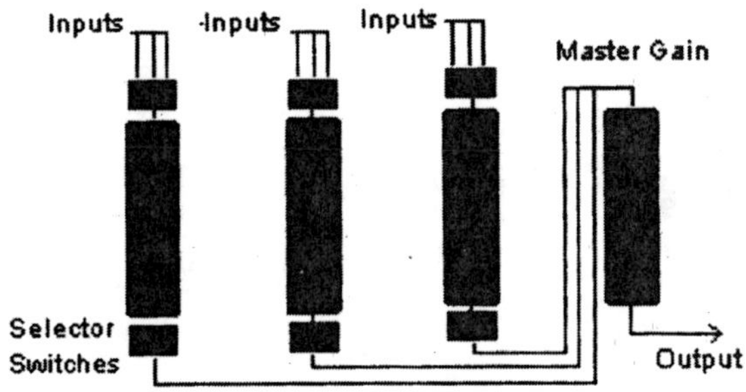

Fig. 3.32. Fader system in a Broadcast Studio.

The entire audio chain is shown in Fig. 3.33. This includes the pre-amp, equalization, and auxiliary sends. If a mixing board has 24 inputs, then there will be equal number of input paths. The most direct path is the sound on its way to the speakers or recording console. The alternate paths allow external effects. For instance, if a digital reverberation unit can be hooked up by taking the output signal from the auxiliary 1, sending to the input of the reverb, process the signal inside the reverb, and then return it via the auxiliary return path. Compressors, limiters, equalizers, enhancers etc. are inserted via channel insert jack. It is common for mixers to "roll-off" excess bass, since the extreme low frequencies contain rumble and wind noise. Low frequency noise can contribute to overload distortion. On interior sets, a mild bass "roll-off", around -6dB at 90 Hz is used. Exteriors often call for a more sever low cut, around -6 dB at 150Hz. Sub-group buses allow grouping of signals in a logical manner. The left and right outputs are the final level control for the signal before it goes to the speakers or recorder. All mixing consoles have these basic functions, and many can be even more complicated. For example, automated console like Yamaha 03D has moving faders and snapshot memory, with facility for sync up to time-code on an ADAT. A 1202 Mixer with 58 knobs is shown in Fig. 3.34.

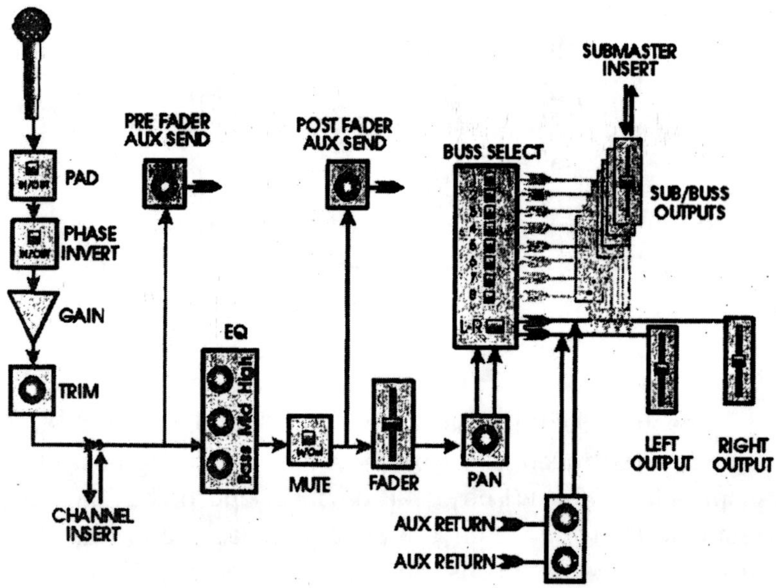

Fig. 3.33. A Typical audio control schematic.

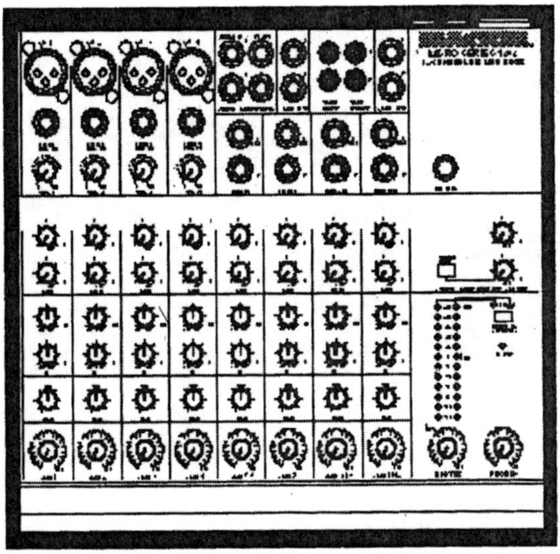

Fig. 3.34. A 1202 Mixer with 58 knobs.

THE VIDEO SWITCHER AND SPECIAL EFFECTS GENERATOR

In a TV studio, the video switcher and special effects generator are used to mix a number of video sources. On the switcher a button called bus represents each video source. The "master" bus controls the final program that goes to air and is known as the "program bus". To produce effects such as fades, wipes, and dissolves, the switcher contains a pair of mix/effects (M/E) busses. To air the final effect, the mix/effects bus is punched up on the program bus. The luminance key triggered by the brightness of the image is used to create a composite image with a camera shot. Chroma key uses colour for its trigger. Many switchers today are digital devices. All the video inputs into the switcher are converted into digital signals before being manipulated. With the use of a digital switcher, hundreds of new effects are available to the user. Few of the effects are: image compression, image expansion, image rotation, mosaic, posterization, negative, monochrome, strobe etc. A typical switcher is shown in Fig. 3.35.

Fig. 3.35. A Typical Switcher.

Television Film Scanning

Two types of equipment have been developed for the transmission of films on television: camera telecines and flying-spot telecines. Both types of telecines have been in use since the early days of television. Each film frame has to be scanned twice to obtain two interlaced fields. For this purpose, the film in a camera telecine is held stationary for the duration of two fields. Fig. 3.36 shows different type of telecines.

- In camera telecines [1], each film frame is focused onto the target of a television camera, via an appropriate lens system.
- In flying spot telecines [2], a light beam, produced on the screen of a cathode-ray tube, scans each film image line by line before activating a photo-electric cell.

With flying spot scanning, the arrangements most commonly used to produce the interlaced scanning pattern are twin lenses [3] and jump scan [4]. In the first case, a shutter covers one of the two lenses for the duration of one frame. With the second, it is the, position of the scanning raster on the cathode-ray tube which, periodically alternates. The use of semi-conductors led to a charge-coupled device (CCD) which resulted in the development of tubeless scanning systems [5]. As discussed earlier in the camera section, an array of photosensitive sensors ("charge-coupled devices") collects the light beam moving through the film and variations in the illumination produce changes in electrical potential at the point of impact of the beam. In this manner the films are scanned for use in TV.

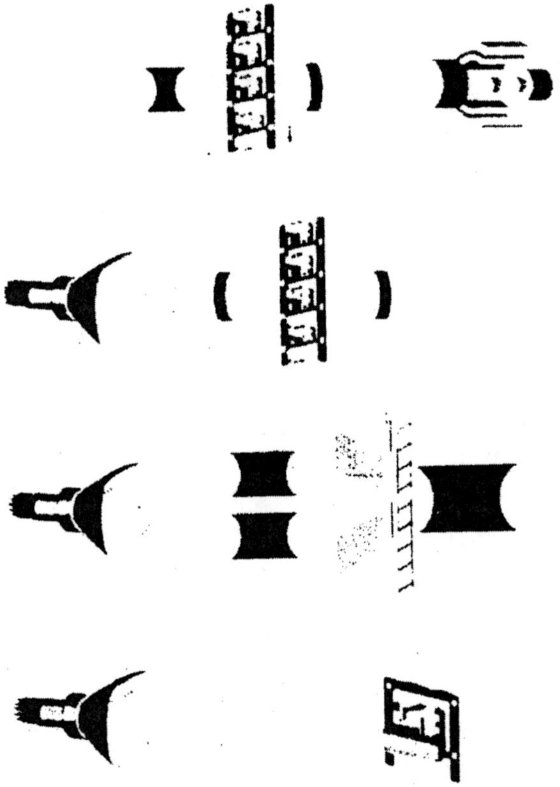

Fig. 3.36. Different Type of Telecines.

Conclusion

In this chapter, the basic studio techniques including production processes, earliest recording methods, microphone and camera characteristics, recording devices, mixing consoles and telecine systems have been described.

REFERENCES

1. The First Sparks: The Birth of Radio at http:// www.cinemedia.com.au/SFCV-RMIT-Annex/rnaughton/ phd4410.html
2. Carter, Alden R., "Radio: From Marconi to The Space Age". at http:// www.cinemedia.nrt/SFCV-RMIT-Annex/mavighton/ MARCONI-1.html.
3. Morse embosser (chain-driven Morse recorder) at http:// www.jct.ac.il/special/virtual/phone/phone1.html
4. Steve Schoenherr, "Recording Technology History: A Chronology with pictures and links" at http://ac.acusd.edu/ History/recording/notes.html#origins
5. The World's First TV Recordings at "http://members.aol.com/ mcleandon/tv_1strx.htm
6. Baird Televisor, 1930 at http://www.bravo.ca/mztv/televisor.html
7. D. F. McLean, "Computer-based analysis and restoration of Baird 30-line television recordings",Journal of the Royal *Television Society*, pp. 87-94, Vol. 22/2 international April 1985.
8. D. F. McLean, "The Recovery of Phonovision, Third International Conference on Image Processing and its Applications", IEE Conference. Publ. 307, pp. 300-304, July 1989.
9. The First TV Recording at http://members.aol.com/mcleandon/ tv_stdio.htm
10. http://www.locationsound.com

4
PRODUCTION TECHNIQUE

"EDITING IS THE CREATIVE FORCE OF FILMIC REALITY...AND THE FOUNDATION OF FILM ART."

V.I. PUDOVKIN, 1915

In Chapter 3, the hardware aspects of the broadcast production system were described. In this Chapter, the production processes and certain techniques used in the real video production shall be discussed. It would be seen that, the production system does not merely consist of shooting the event or picking up the voice and blending the two. This is rather a creative process and much effort and artistic venture has to go through if a dramatic effect has to be created. The effect of voice and picture on human mind has tremendous capacity to capitulate it. The process of production is the transfer of one's own mind to the viewer /listener and hypnotize him[1,2]

BROADCAST PROCESS

The broadcast process consists of initial recording/shooting, post-production and distribution. Initial production involves shooting in studios, OB (outside broadcast) coverage, etc. Post-production consists of editing, layering, modification of picture contents and adding audio/video effects while distribution consists of actual delivery via terrestrial transmitter, satellite or cable (Fig. 4.1). Certain pre-production processes start for every video and film before the production process. Thus, in the studios, the sequence is pre-production, production and post-production, which may be referred as scripting, shooting and editing. Different type of shots viz. close-ups, action shots and quick cutting are used to produce scenes. A film or video production also includes audio, while sound broadcasting uses audio alone.

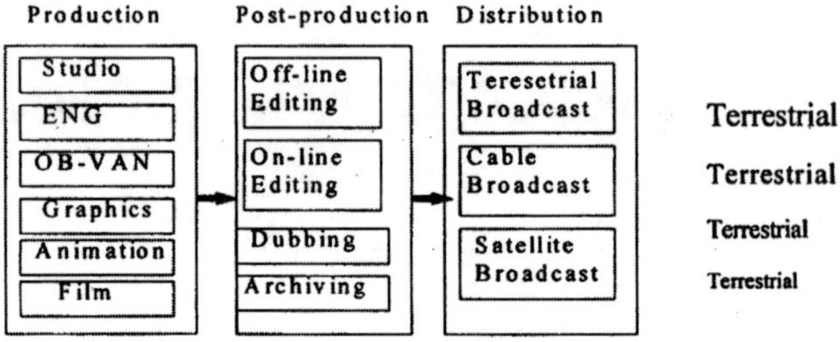

Fig. 4.1. Broadcast Processes.

All videos and films are created from the same elements like close-up shots, titles, music and narration. When skillfully combined, these visual and audio elements provide a powerful impact on the viewer. But before the discussion of how it is done, it is worthwhile to know who does what in film/video making.

ACTORS IN THE SCENE

The person who is generally in charge of the entire production is the producer. Producer comes up with the program concept, lays out the budget for the production, makes major decisions, works with the writers, decides the key talent, hires the director, and guides the general direction of the production. Writer (also called scriptwriter) provides a script that tells everyone what to do, say, etc.). Actors, reporters, hosts, guests, and off-camera narrators—anyone whose voice is heard or who appears on camera comes under the talent category. Thus, talent includes actors (who portray other people in dramatic productions), performers (who appear on camera in non-dramatic roles), and announcers (who typically do not appear on camera). The Director is in charge of working out pre-production details, coordinating the activities of the production staff and actors, working out camera and talent positions, selecting the camera shots during the production, and supervising post-production work. Technical director is responsible for coordinating the technical aspects of the production. A lighting engineer designs the lighting plan, arranges for the lighting equipment needed and sets up and checks the lighting. A set designer designs the set and supervises its construction, painting and installation. A make-up person is responsible for make-up, hair spray, etc. There may be other persons like associate producer, production assistant, cameraman, engineers to assist in the production[3,4,5].

VIDEO SCRIPT

The script is a blueprint for creating an effective film of any kind. The script indicates the type of shots to be taken at various stages. This provides a guideline to the director and cameraman for shooting the film the writer has envisioned. This also provides a blueprint to the editor for editing the film. But more importantly, the script is a preview of the film on paper. It is a lot easier and cheaper to change a script on paper than it is to change a film or video.

THE CAMERA SHOTS

CAMERA FRAMING

The framing of a shot, at a most rudimentary level, determines what we can and cannot see.

WIDE SHOT

A wide shot also called the establishing shot, shows the subject(s) in their environment. For instance, shot #1might show most of a cricket field with stadium and the background. In another scene, it could show the ocean, the person, etc. Fig. 4.2.

Fig. 4.2. Wide Shot.

MEDIUM SHOT

This shot is closer than the wide shot. When applied to people, it shows the subject(s) from waist to just above the head. In comedy the use of medium shots keeps the mood light. Fig. 4.3 and Fig. 4.4

Fig. 4.3. Medium Shot.

Fig. 4.4. Medium Shot.

CLOSE-UP

The actor is framed from his or her chest to just above his or her head. Close-ups on individuals are appropriate for interviews and dramas. A close-up of a person may show the head only Fig. 4.5. In a news presentation, when the newscaster speaks directly to the camera, a close up shot may be only the "talking head."

Fig. 4.5. A Close-up Shot.

SHADOW SHOT

A back lit shot can actually be an artistic technique when used properly. It provides a dramatic effect Fig. 4.6 6.

Fig. 4.6. Shadow Shot.

OTHER CAMERA SHOTS

A few other camera shots are:

Dolly shot

The camera is placed on a moving platform (dolly) or vehicle to follow along a moving subject so the camera moves over ground rather than panning from a fixed tripod.

Zoom shot

The camera lens zooms in or out.

Extreme Close-up

It may show eyeballs and mouth Fig. 4.7. When applied to inanimate objects, such as a dime, the object may fill the frame. Extreme close-up provides great impact and value.

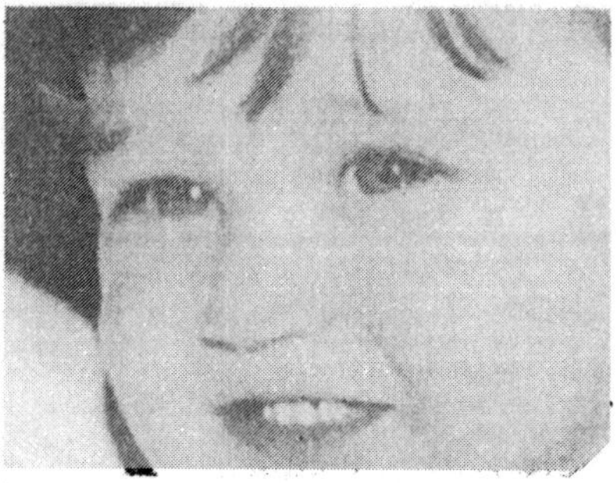

Fig. 4.7. Extreme Close-up.

Pan left/right

The camera moves along a horizontal path to follow movement from the left or right. This shot is also used to reveal the full view of a wide object like a train or the Grand Canyon.

Tilt up/down

The camera moves along a vertical path. Like the pan, this shot allows the viewer to follow movement like a rocket launch Fig. 4.8 or to see all of a tall object like a skyscraper.

Fig. 4.8. Tilt up /down Shot.

Aerial

A moving camera shot like a dolly shot made from a helicopter or airplane Fig. 4.9.

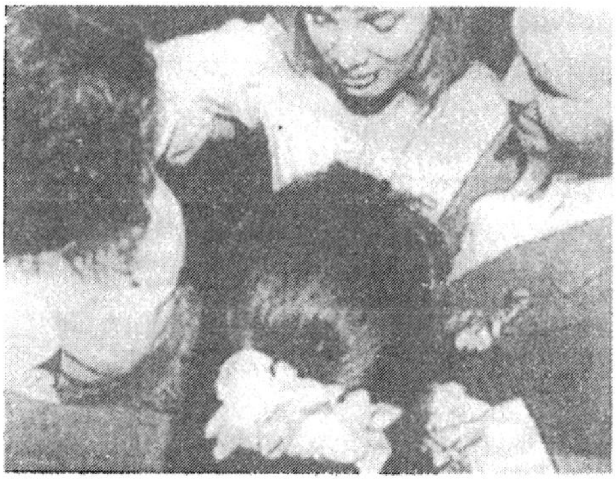

Fig. 4.9. Aerial / Overhead Shot.

A significant amount of picture detail is lost in the 525 and 625 line broadcast television systems. The only way to show details is through close-ups. Except for establishing shots designed to momentarily orient the audience to subject placement, the director and the editor should emphasize medium shots and close-ups[7,8].

COMPUTER GENERATED TEXT AND GRAPHICS

Text and graphics on television are created with computers rather than filmed with the camera[9,10,11]. There are several techniques.

SUPERIMPOSED TEXT AND GRAPHICS

During an interview, the title of the interviewee is superimposed over the lower third portion of the screen on a medium shot. This is called "a lower third." Many times a line is superimposed over a video shot or still photo.

ANIMATION

Flying network logos, a spinning globe that opens a news show are examples of animation. Animation can standalone or be superimposed over the live video shots. Today these are all created on computers.

TRANSITIONS

Transitions are used to join the shots together. The order in which the shots are cut together and their length is one of the essential arts of editing. Four types of transition generally used are straight cut, fade, dissolve or wipe. In a script, all transitions from one shot to the next are assumed to be straight cuts unless otherwise indicated.

The straight cut

Straight cut simply butts the end of one shot with the beginning of the next. 99 per cent of the joints are made with the straight cuts. The straight cut is used in a wide range of films from the simplest informational film to the most complex feature film.

Fade to and from Black

Most commercials actually begin with a black screen. The first shot fades up from this black screen. The duration of this fade varies from a fraction of a second to as long as two seconds. The shot becomes progressively brighter until it is as bright as the shots that follow it. "Fade up from black" is a visual clue to the viewer that something new is happening. It may be the beginning of the film or the beginning of a new section of the film. It can indicate to the viewer that a period of time has elapsed from the last section of the film. This lapse can be of few minutes or years.

Dissolve

The dissolve is a transition that uses two fades. Shot #1 fades down to black. Shot #2 fades up from black. The difference is that there is no black. As shot 1 is fading down, shot 2 is already fading up. The fading parts of these shots are superimposed over each other. Like fading to or from black, the dissolve indicates to the viewer a different time or place, with more sensation.

Wipe

There are many types of wipes. A simple one is where the end of shot 1 moves off the screen to the left while the beginning of shot 2 moves in from the right until it fills the screen. Shot 2 appears to push the first shot off the screen. Another is the circle wipe. The end of shot 1 develops a hole in the centre that grows larger. Inside the hole in shot 1 is the beginning of shot 2. This circle finally fills the frame until shot 2 has completely replaced shot 1. There are a number of other geometric patterns where shot 1 is replaced by shot 2. These include a vertical line, horizontal line, boxes, triangles, and more. Wipes, like fades to black and dissolves, may indicate a change in time or place.

MIXING OF AUDIO

The sound track, which is often music and dialogue, helps the shots flow smoothly into one entity, but even more significantly, the sound track often transcends the impact of the visual elements, leading the viewer to experience the film or video on an emotional level.

There are 4 basic types of audio: music, narration, synch sound and sound effects.

MUSIC

Different types of music, accompanying the same visuals, can dramatically change the mood and impact of the video. Many videos and most commercials start with music and this music fades up. The same music may continue throughout, or it may cut or dissolve to another piece of music. Commercials use rock 'n roll, classical, country, opera and every other type of music. Many of the music selections used in commercials have a lively beat, which helps drive the visuals and makes the film more compelling. It may be purely instrumental or have lyrics sung by a vocalist.

Some scriptwriters specify exactly which song is to be edited into their commercial. Others only describe a mood—upbeat, contemporary, and happy—and let the editor find just the right music.

NARRATION (VOICE OVER)

Narration or voice over (VO) is provided by a male, female, young or old along with the picture but this is done behind the scene. The script often describes the character of the voice, e.g., helpful young man, wise grandfather, the voice of authority. Sometimes the narrator may be talking straight in front of a background, while the camera may cut to other scenes he describes, but his voice continues over the visuals. The scene may end by cutting back to him on-camera as he continues talking. In scripts these are referred to as an on-camera narrator and an off-camera narrator or Voice Over (VO). In many scripts, there may be more than one narrator or even a dialogue.

SYNCH SOUND

Synch sound is sound is synchronized with the picture. In this case camcorder records sound and picture simultaneously (in synch). Bigger budget commercials are shot with a camera which records only the picture and a separate tape recorder records sound. Before editing begins, these separate pieces of film, one with the pictures and the other with the sound track must be synchronized so that the actors' lips move exactly like when he speaks. Thus, the picture and sound track are in synch.

SOUND EFFECTS

Many sounds in a film appear to be synch sounds that are recorded later in a special studio. Gun shots are a good example. Another common sound effect is footsteps. Foley artists create these enhanced footstep sounds (After Jack Foley, a sound editor). These sound effects had long been used in radio shows. In film these effects had to be exactly synchronized with the picture.

SELECTING A MICROPHONE

Microphone selection plays an important role in overhead miking technique. There is no one choice of specific microphone that will be right for all situations. A professional package should include an assortment. A majority of the production dialogue recorded in theatrical productions and television series is miked from overhead, utilizing either a fishpole or studio boom. Overhead miking provides a natural sound. Normal sound effects and some background ambience are also picked up, at lower, relative level than the dialogue, thus rounding out the total track. Perspective matches camera angles, since the boom mike is able to get in closer on tight shots, and is further away on longer angles.

The highest quality condenser microphones are used in capturing dialogue. Top of the line condensers offer superior reach and sensitivity, and that can spell out the difference on those more demanding shots between getting 'rich' dialogue versus 'weak' or 'thin' audio. Lavalieres and radio mikes are not used for TV productions since these lack in picking natural sound effects and ambience. Further, audio picked up by lavaliere is often subject to abrupt changes due to talent turning their head off-axis, or leaning over a hard surface (such as a desk or podium). Lavalieres are also prone to distracting clothing noise and other interference. Similar to telephoto lenses, shotgun microphones tend to compress the distance between foreground and background. The best way to eliminate this "telephoto" effect is to aim the mike down from above, so that the only "background" in the microphone line of sight is the silent ground.

The most popular microphones for exterior use are long shotgun microphones. Long shotgun microphones offer narrow pick-up patterns and excellent sensitivity and reach. Deployed overhead of talent, tilted just slightly towards the mouth—these mikes eliminate considerable background ambience while still picking up natural sound effects such as footsteps. Because of their directionality, these long shotguns can be played at greater headroom above the actor when necessary (up to several feet depending on ambient noise), thus facilitating the wider frames more common on exterior set-ups.

The disadvantages of the long shotgun are its directionality and physical dimensions. The narrow pick-up pattern requires that much more care is to be taken in cueing (aiming). Moving talent must be meticulously followed; multiple talent requires rapid and precise repositioning of the mike for each person's lines. A compromise could be a short shotgun with more manageable length and wider pick-up pattern (supercardioid or hypercardioid).

Shotgun microphones with interference tubes tend to exhibit loss of definition when used in confined, hare-walled interiors such as kitchens, bathrooms, small offices, and so on. This phenomenon is caused by reflected sound waves interfering with the acoustic noise. The solution is to use the wider-angle condenser mikes without interference tubes. The problem is that these wider-angle mikes (cardioids and hypercardioid) also have less reach. It becomes a matter of trade off between choosing the crisp sound offered by the wider pattern mikes versus getting the reach needed to isolate dialogue from the background noise due to higher distance often demanded by the camera angle.

When recording dynamic scenes, in which the actor is using a hand-held microphone as a prop, it is better in most cases not to take the dialogue feed from the same mike. Actors may use their hand mike as objects to accentuate their visual performance, without regard for what their broad gesturing and handling may do to the sound. It is better to pick sound of "stage performers" either from overhead from a fishpole or boom, or off a concealed lavalier.

Whether held in the hand or mounted on a stand, the microphone should be positioned about 6-12 inches from the talker's mouth, pointing up at about a 45^0 angle. With some types of microphones, holding the microphone very close (3-6 inches) causes additional emphasis of the lower frequencies (known as proximity effect), resulting in a "warmer," bass-heavy sound[12,13,14].

SOUND TRANSITIONS

The same transitions that join visuals are used to join the audio cuts. It may be noticed that in many transitions, the music fades down slightly as the narrator's voice begins. The music may fade out at the end of the commercial. The bird chirp is probably a straight cut. Some sounds overlap just like a dissolve overlaps two shots. The audio or sound track can have a remarkable effect on the visuals. Many sound tracks of commercials can stand alone without visuals.

The shots and sounds flow together in a sequence that tells a little story about one thing. Since commercials are so short, they may only have one such sequence, but longer films will have many sequences. They are called scenes. Each scene tells a small story that is part of a bigger story. They flow logically one from another, each building on the last.

VIDEO TAPE EDITING

Editing allows for the creation of a story by cutting unwanted footage out of production and inserting effects, stored scenes, captions, sound, etc. Editing is a powerful tool and the possibilities for a production are endless with its effective use. It is the job of the video editor to cut the footage down without leaving out anything important, and without abrupt changes in mood, pacing, energy, or rhetorical direction.

NON-LINEAR EDITING

The limitation of analog technology with regard to the number of copies and layers are not the only drawbacks. The tape recorder needs time to play, record, wind and rewind the tape to find the desired clip. The digital video/audio server and disk recording technology provide random access, ease of editing and post-production. As such the trend is towards a digital production system.

TYPES OF EDITS

There are essentially two types of editing, assemble (continuity) editing and insert editing. Before describing these two edits, it is worthwhile to recall how the VCR records the signals onto tape. Let us consider a pre-recorded tape with two audio tracks, one video track, and one control track. The audio and video tracks record the audio and video respectively. The control track keeps the tape running at an appropriate speed and keeps the picture stable on-screen. When editing, these tracks become altered.

Assemble Editing

Assembly or continuity editing refers to arranging the sequence of shots to suggest a progression of events. It is very convenient to record in assemble mode, because each segment can be pieced together efficiently and quickly. Given the same shots an editor can suggest many different scenarios. In assemble mode, the editing VTR records a new control track, audio tracks, and video track.

Insert Editing

Insert edit inserts video and audio over a stable control track of a "pre-blacked" tape. This basically means that the tape already has the video black recorded from a blackburst generator. With insert editing, it is possible to insert any combination of video and audio.

EDIT CONTROL SYSTEMS

There are three types of edit control systems viz. control track editing, time-code editing, and computer editing.

Control Track Editing

Using the control track pulses on the videotape, the edit controller tracks the number of frames on the tape. The problem with control track editing is the lack of an absolute frame marker on the tape.

Time-Code Editing

A time-coded tape correctly shows the address on edit controller. This is very valuable since instead of rewinding each tape to the beginning and resetting the counter to zero, the counter automatically adjusts itself.

There are several ways of recording time-code on tape:

Vertical Interval *Time Code* (VITC)

Time code signals are recorded in the vertical interval on the tape eliminating the need for a separate audio track. The problem of VITC is that it needs to be recorded at the same time the video was recorded (i.e. the camera must have VITC capabilities).

Longitudinal *Time Code* (LTC)

Time code signals can be recorded on a separate audio track if available. The only problem is that the time code tones may sometimes "bleed" over to the other audio track if the level is not properly adjusted.

RC Time Code (Rewritable-Consumer *Time Code*)

This *time code* is recorded on 8 mm and Hi-8 tapes. There is a dedicated track for the *time code*, freeing both audio tracks.

Control Track Time Code (CTL)

Introduced by JVC Corporation, it records *time code* on the control track but it is only compatible with their equipment (S-VHS).

The great part about *time code* editing is that the producer or videographer can make a "burn-in" dub of the time coded tapes onto VHS to view at home or any other VCR that does not support *time code*. A burned-in dub means a window that contains TC information which is displayed on the TV image. A written edit decision list can be made using these numbers in the window.

Computer Editing

The final editing system is computer editing. Using the time code capabilities of the editing VCRs, the computer can store information regarding any list of edits into an EDL or edit decision list file. A great product called the FutureStation V-Link connects a PC to editing VCRs and acts like an editing controller. There are also other products available for the Amiga and Macintosh.

A/B Roll Editing

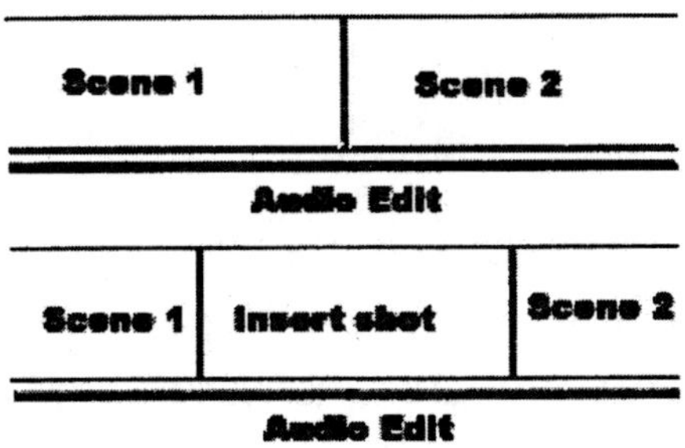

Fig.4.10. A/B Roll Editing.

The videotape used to record the main event is known as A-roll. Another videotape known as B-roll is used to record cutaway shots viz. reaction of the audience etc. Fig. 4.10. In linear editing this process of recording on two separate videotapes makes editing much easier. While cutting a section out of the scene insertion of a three- or four-second cutaway shot eliminates an abrupt and noticeable jump in the video.

Cuts-only edit can be performed with the help of two VCRs. In order to introduce effects viz. dissolves, wipes, chroma-keys, graphics, etc., a switcher or special effects generator is required. The computer editing systems allows the videographer to integrate the system with an external switcher.

In order to produce a dissolve from one scene to another. First, a dub of the playback tape is made. Thereafter another playback VCR is used to perform the edit. This is called A/B roll editing. Using the computer, the time of dissolve between the two scenes can be decided automatically. There should be absolutely no shift in hue, brightness, or resolution between the A and B rolls.

TIME BASE CORRECTOR

All recordings on analog, mechanical video equipment are subject to time base errors resulting due to variations in tape speed. The errors appear on the screen as picture jitter, "skewing" (shifting of the video image), or color break-up. These errors are corrected by feeding the signal to a TBC, where each frame is digitized and corrected for errors. The TBC performs the correction by

- realigning the video image to prevent skewing
- allow the editor to adjust the saturation (color or chroma), brightness (luma), black level (a.k.a. pedestal or set-up), hue (tint) and timing characteristics of the video signal
- By dropout-compensator (DOC) that "fills-in" the white streaks appearing in a video signal played back from a worn-out or dirty tape or from a VTR with dirty heads.

Many digital VTRs based on the popular DVCPRO, DVCAM, and Digital-S formats include some form of time base correction.

CONCLUSION

In this chapter, the production techniques consisting of shooting, audio mixing and video editing have been described.

REFERENCES

1. Aldridge, Henry B. and Lucy A. Liggett. "Audio/Video Production: Theory and Practice", Englewood Cliffs, NJ: Prentice-Hall, 1990.

2. Armer, Alan A. "Directing Television and Film". 2nd ed. Belmont, CA: Wadsworth, 1990.

3. Barr, Tony, " Acting for the Camera", New York: Harper and Row, 1982.

4. Blumenthal, Howard J., "Television Producing and Directing", New York: Barnes and Noble, 1987.

5. Burrows, Thomas D. et al., "Television Production: Disciplines and Techniques'" 5th ed. Dubuque, IA: Wm. C. Brown, 1992.

6. http://www.elitevideo.com/story

7. Kenney, Ritch, and Kevin Groome, "Television Camera Operation," Burbank, California.: Tellem Publications, 1987.

8. Millerson, Gerald, "Video Camera Techniques", 2nd ed. Newton, MA: Focal Press, 1994.

9. McQuillin, Lon, "Computers in Video Production," White Plains, N.Y.: Knowledge Industry Publications, 1986.

10. Wershing, Stephen and Paul Singer, "Computer Graphics and Animation for Corporate Video". White Plains, NY: Knowledge Industry Publications, 1988.

11. Weynand, Diana, "Computerized Videotape Editing", Woodland Hills, CA: Weynand Associates, 1983.

12. "Location Sound Corporation" at http://www.locationsound.com

13. "From Sound Wave - The Production Sound Newsletter" at http:\\www.taiaudio.com

14. http:\\www.pro-sound.com

5
TRANSMISSION TECHNIQUE

"WIRELESS (IN THE 1920s AND 30s) WAS A WONDER. CELEBRATED AS A NEW SCIENCE FOR THE UNIVERSAL BENEFIT OF HUMANITY."
LESLEY JOHNSON, THE UNSEEN VOICE 1988

INTRODUCTION

The transmission technique of radio communication has undergone a complete change from the earliest spark transmitter which was use by marconi, to the present digital system. The broadcasting systems of the future will use digital data transmission systems that would include images. Progress has been made in research and development on the use of digital broadcasting technology to integrate a number of broadcasting services into a single digital broadcasting wave, thereby ensuring ISDB (Integrated Services Digital Broadcasting) as a new broadcasting system to provide advanced multimedia services. In this part, the transmission theory and the modulation methods as well as the new digital broadcasting methods have been described.

RADIO WAVES

Radio waves are part of a general class of waves known as electromagnetic waves. In essence, they are electrical and magnetic energies that travels through space in the form of a wave. They are different from sound waves which are pressure waves that travel through air or water. Electromagnetic waves at certain frequencies show up as radio waves. At much higher frequencies, they are called infrared light. Still higher frequencies make up the spectrum known as visible light. This goes on up into ultraviolet light, and X-rays. Radio waves have two important characteristics that change. These are amplitude and frequency.

PURPOSE OF TRANSMITTER

The prime function of the transmitter is to generate the RF carrier frequency to be transmitted. Two ways in which this carrier is generated is the use of a crystal oscillator, or the use of frequency synthesis (phase locked loops). In most modern transmitters it is the phase lock loop that generates the carrier. This generated carrier will is modulated, amplified, and then transmitted through the airwaves by use of an antenna network.

FREQUENCY GENERATION USING CRYSTAL OSCILLATORS

An amplifier generates the RF signal with the crystal in the feedback path, causing the amplifier to oscillate at the crystal resonant frequency. Crystals are restricted to the frequency range of 1 MHz to 20 MHz. Below 1 MHz the crystal element becomes too large for most applications. Above 20 MHz the crystal element is very thin and fragile.

MODULATION OF THE CARRIER

The purpose of modulation is to modulate the carrier with the desired information to be transmitted. For amplitude modulation, the modulation is accomplished in the final power amplification stages. For frequency modulated systems, it is done usually at the second oscillator if it is not synthesized. If the system is using a synthesized FM modulation, then phase modulation after the first oscillator is used.

AMPLIFICATION

Non-synthesized transmitters (directly using the crystal as the source) need to multiply the signal as well as increase its' amplitude. The two functions are performed with the use of medium power amplifiers. The amplifiers are arranged in a chain, with each stage having more gain than the stage before. The first few stages are operated in the non-linear region to generate harmonics. With FM synthesized transmitters, output of the 2nd oscillator is amplified, and this oscillator runs on the correct transmitter frequency.

Once the signal is at the correct frequency, amplified to desired level, and modulated with information to be transmitted, it is applied to an antenna without losing any power. The signal is routed through a matching circuit to bring the impedance to the value necessary (typically 50 ohms for mobile radio). The signal is routed through the antenna switch to a harmonic filter and then to the antenna.

EARLIER TRANSMITTERS

SPARK TRANSMITTERS

The transmitter is the heart of every radio station. If the transmitter stops working, the radio station is "Of the Air". The earliest transmitters, including the one used by Marconi employed spark technology. A spark transmitter emitted a series of damped waves, the period of which was the spark rate.

A spark transmitter consisted of a spark gap connected across an oscillatory circuit comprising of a capacitor and an inductor in series (Fig. 5.1). The capacitor was charged to a high voltage by an

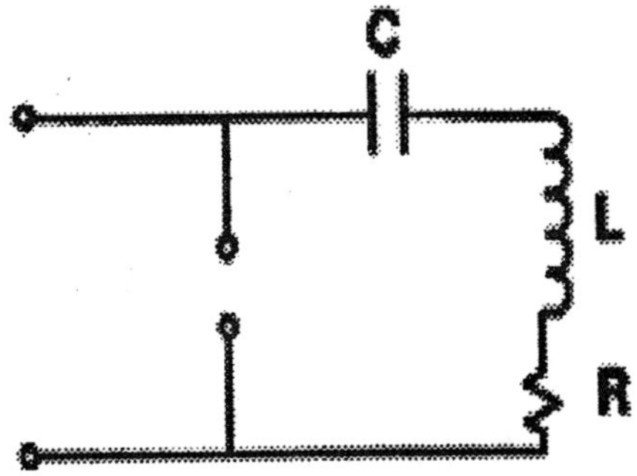

Fig.5.1. Basic Spark Transmitter.

induction coil. When the potential across it was sufficiently high to breakdown the insulation of air in the gap, a spark was generated.

Since this spark had a comparatively low resistance of the order of one or two ohms, the spark discharge was equivalent to the closing of an L-C-R circuit. The condenser then discharged through the conducting spark and the discharge took the form of a damped oscillation, at a frequency determined by the resonant frequency of the spark transmitter. The circuit for the spark transmitter used by Marconi is shown in Fig. 5.2.

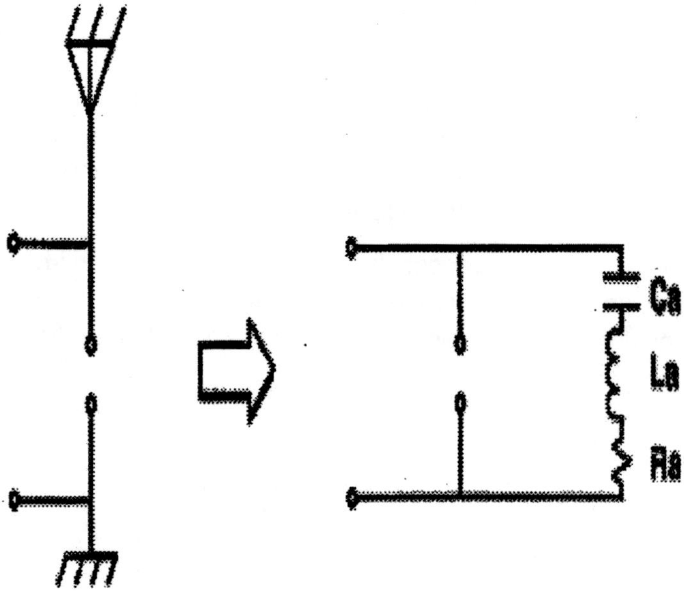

Fig. 5.2. Circuit of Spark Transmitter Used by Marconi.

Earlier spark transmitters employed very high capacitance condensers, to resonate with a 1 or 2 turn inductor, in order to achieve high transmitter powers Fig. 5.3.

Fig.5.3. Marconi's Clifton, Ireland condenser under construction after Guglielmo Marconi, published by the Marconi Company Ltd., 1984.

Marconi's first transmission in Morse code was: CQ, ("seek you"), calling any station or calling all stations, from PN. PN was the call sign used by the Marconi station at Poldhu, Cornwall. He used an end-fed wire aerial (a monopole).

Spark transmitters were used from about 1900 to about 1925[1]. Efficiencies for early spark transmitters were about 60 per cent. A radio transmitter with an aerial power of 300 watts, used by Marconi in 1926 which remained in operation up to 1931is shown in Fig. 5.4. The transmitter nicknamed "Lausanne-Champ-de-l'Air" is on display in the telecommunication pavilion of the Museum of Transport and Communications in Lucerne. The first radio concert in Switzerland was broadcast from this transmitter on 26 October 1922[2].

Fig. 5.4. Marconi radio transmitter Lausanne-Champ-de-l'Air (1926-1931).

Source : http://www.jct.ac.il/special/virtual/phone/phone11.html

ALTERNATOR BASED RADIO TRANSMITTER

The next in the series of transmitter development was an alternator based radio transmitter. In its simplest form, an alternator consists of a magnet, called the field, and a piece of conductor known as armature. Any time the armature is in the influence of a magnetic field that varies in strength, an electric current is set up in the conductor. The strength of the magnetic field can be varied in a number of ways. One method to create relative motion between the field and armature is by forming the armature into a loop and rotating it in the field. The same results are produced if the armature is held stationary and the field is rotated. In some designs, called inductor alternators, the variations in magnetic strength are produced by moving a piece of magnetic material between the field and the armature. All of these techniques were used in radio alternators.

The voltage at the output terminals of an alternator has a frequency, which is determined by the design of the machine and the speed at which it is operated. In theory, if any alternator is operated fast enough, its output will be high enough in frequency to be used directly for radio.

Typically, the initial frequency produced by the rotating armature in a Goldschmidt alternator was 15,000 cycles per second. A tuning circuit of the same frequency, connected to the armature caused that energy to be reflected to the stationary field 30,000 times per second. Another tuned circuit at 30,000 cycles per second connected to the field then built up the energy at that frequency and re-induced it back into the armature, where its frequency was added to that of the armature, bringing it to 45,000 cycles per second. It was then reflected one last time back to the field with another 15,000 cycle increase to 60,000 cycles, at which frequency it was radiated from the antenna[3].

Efficiencies of 80 per cent were claimed for the Goldschmidt's alternator based transmitter, however, generally alternator based radio transmitters operated at overall efficiency of 66 per cent, and maximum output power of 500 kW.

VACUUM TUBE TRANSMITTERS

The three-element vacuum tube was well known by 1915 to be capable of regeneration and oscillation. Various types of modulation became possible with the development of vacuum tubes.

MODULATION

The information signal can rarely be transmitted in its original form. For electromagnetic transmission, it must first be converted from audio into an electric signal. A transducer accomplishes the conversion. After conversion it is used to modulate a carrier signal. A carrier signal is used for two reasons:

- To reduce the wavelength for efficient transmission and reception. The optimum antenna size is ½ or ¼ of a wavelength. A typical audio frequency of 3 KHz has a wavelength of 100 km and would need an effective antenna length of 25 km. By comparison, a typical carrier for FM is 100 MHz, with a wavelength of 3 mm, and could use an antenna of only 80 cm long.

To allow simultaneous use of the same channel called multiplexing. Each unique signal can be assigned a different carrier frequency and still share the same channel.

The process of modulation involves systematic use of the information signal to vary some parameter of the carrier signal. The carrier signal is usually just a simple, single-frequency sinusoid (varying in time like a sine wave).

The basic sine wave is represented by the following formula

$$V(t) = V_o \sin(2 \pi f t + \phi)$$

where,.

- $V(t)$ represents the voltage of the signal as a function of time.
- V_o represents the amplitude of the signal (the peak value achieved each cycle).
- f represents the frequency of oscillation, the number of cycles per second (Hertz = 1 cycle per second).
- ϕ represents the phase of the signal.

To modulate the signal, one of the three parameters of the signal viz. the amplitude, frequency or phase is systematically varied. Therefore, the type of modulation may be categorized as either amplitude modulation (AM), frequency modulation (FM), or phase modulation (PM).

AMPLITUDE MODULATION (AM)

AM was invented in early 1900s and popularized by De Forest. Using AM technique, commercial radio station began broadcasting in Pittsburgh in 1920[4]. The principle of working of a typical AM transmission system is shown in Fig. 5.5. It would be seen that the information signal is mixed with the carrier signal to produce the full AM signal to be transmitted.

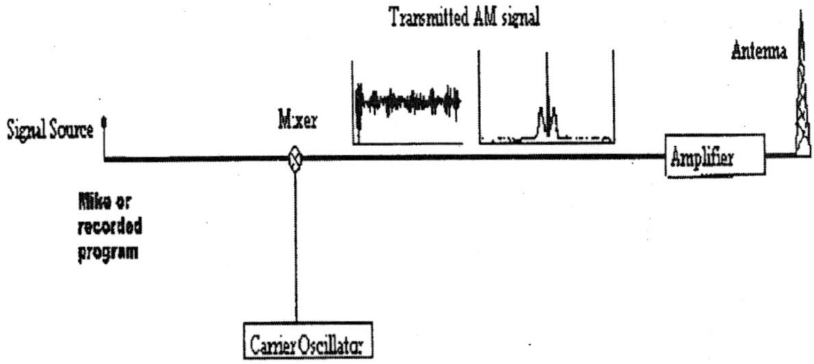

Fig. 5.5. Principle of AM Transmission System.

In Amplitude, the transmitter uses the information signal, $V_m(t)$ to vary the amplitude of the carrier, V_{co} to produce a modulated signal, $V_{AM}(t)$. The three signals in mathematical form are given below:

Information: $V_m(t)$

Carrier: $V_c(t) = V_{co} \sin(2\pi f_c t + \phi)$

AM: $V_{AM}(t) = \{V_{co} + V_m(t)\} \sin(2\pi f_c t + \phi)$

From the above, it would be seen that the amplitude term has been replaced with the combination of the original amplitude plus the information signal. A typical AM signal, showing the carrier signal modulated by the information signal is shown in Fig. 5.6. and 5.7.

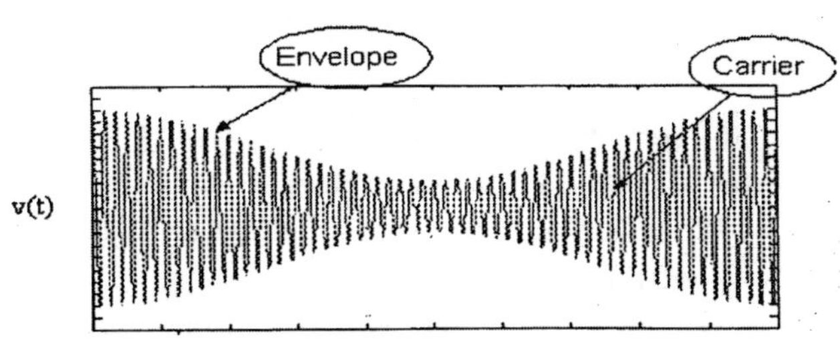

Fig. 5.6. A Typical AM Signal.

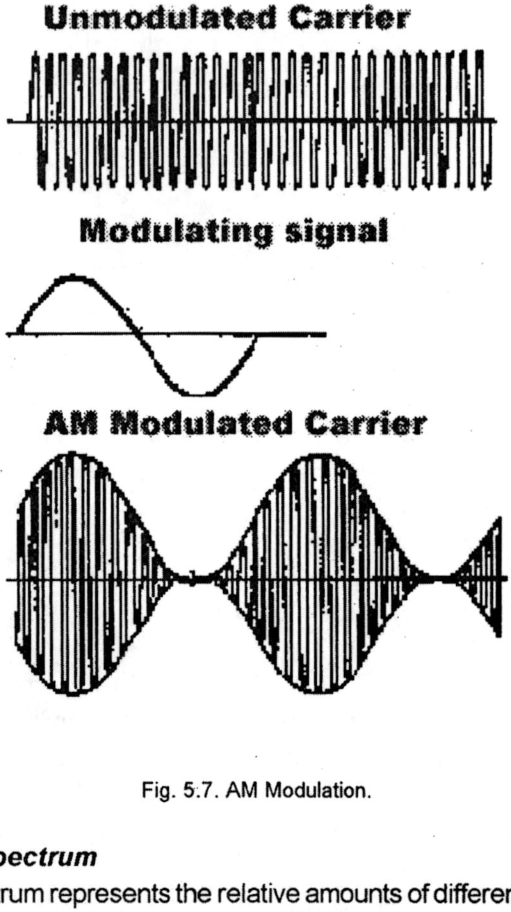

Fig. 5.7. AM Modulation.

The AM Spectrum

A spectrum represents the relative amounts of different frequency components in any signal. A signal can be decomposed into purely sinusoidal components with a few pathological exceptions. In technical terms, the sines and co-sines form a complete set of functions, also known as a basis in the infinite-dimensional vector space of real-valued functions (gag reflex). Given that any signal can be thought to be made up of sinusoidal signals, the spectrum represents the composition of sinusoids viz. one part of 50 Hz and two parts of 200 Hz. Pure sinusoids have just one component. Fig. 5.8 shows a carrier of 8 Hz with a spectrum consisting of a single component with value 1.0 at 8 Hz. Fig. 5.9 depicts the spectrum of amplitude modulated signal for a pure sinusoidal information signal. In this case, the carrier has a frequency of 65 Hz, and the modulating signal is at 5 Hz. The modulation index is 0.5. The spectrum reveals that there are two side-bands on either side of the carrier.

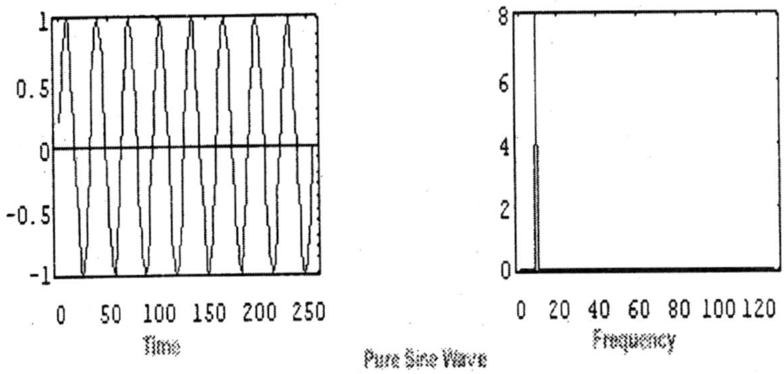

Fig. 5.8. Spectrum of a Single Frequency Carrier.

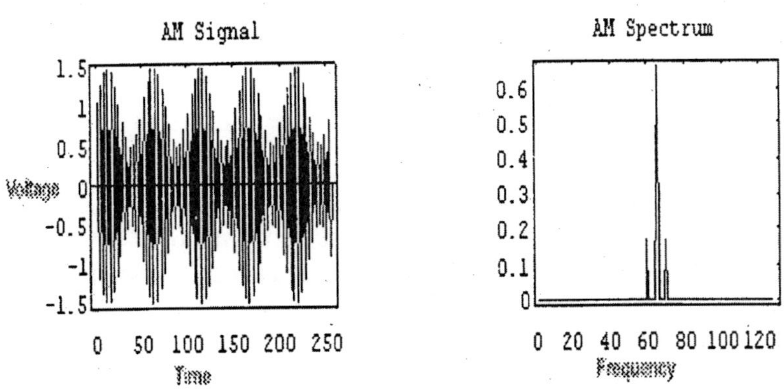

Fig. 5.9. Spectrum of Amplitude Modulated Signal for a Pure Sinusoidal Information Signal.

The information signal is rarely a pure sinusoid. In a real situation, it is a complex mixture of pure sinusoids. As already mentioned, any signal can be decomposed into its pure sine and co-sine functions. This complexity is reflected in the spectrum of the AM signal used to transmit the music to radio. Spectrum of a typical signal is shown in Fig. 5.10.

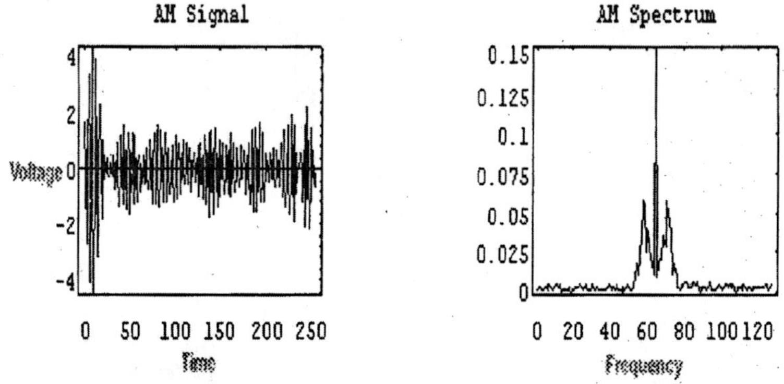

Fig. 5.10. Spectrum of typical amplitude modulated signal.

Bandwidth

Bandwidth (BW) is the measure the range of frequencies that the entire signal occupies. It is represented by the formula

$$BW = 2f_m$$

where, f_m is the maximum modulating frequency.

The bandwidth of a signal determines the number of channels (or stations) that can be operated in a specific band without interference. Higher the bandwidth, higher is the fidelity but lower is the number of channels.

Efficiency

The efficiency η, of the modulation is expressed by the formula:

$$\eta = P_{SB} / P_{TOT}$$

where, P_{SB} = the power in all the side-bands

P_{TOT} = the total transmitted power (includes carrier and side-bands).

Modulation Index

The amount of modulation depends on the amplitude of the information signal. This is called modulation index and expressed as a ratio of the maximum information signal to the amplitude of the carrier.

Modulation Index, $m = MAX(V_m(t))/V_{co.}$

If the information signal is also a simple sine wave the modulation index has a simple form:

$$m = V_{mo}/V_{co}$$

Thus, the modulation index, m is the measure of modulation. Higher modulation results in more power into the side-bands which is the desirable feature. If the modulation index is increased, the efficiency also increases. Accordingly, largest value of modulation index, m should be used. However, if m is higher than one, a distortion is introduced in the received signal.

AM Radic

AM radio is the most common example of amplitude modulation. The frequency band used for AM radio is 531 to 1602 KHz in MF band. The information transmitted is music and voice which fall in the audio spectrum. The full audio spectrum ranges up to 20 KHz, but AM radio limits the upper modulating frequency to 5 KHz. This results in a maximum bandwidth of 10 KHz. Therefore, the stations can be assigned frequencies that are 10 KHz apart without fear of overlap. However, under ITU plan, the channel separation was made 9 KHz. As such, roughly 111 transmitter frequencies are available in this band.

The limiting of the upper frequency removes some of the original information. Since the ability to exactly reproduce the signal is called fidelity, there is a loss of fidelity in AM broadcasts. This is one of the reasons that AM radio does not sound as good as FM radio.

The efficiency of AM transmission is limited by the restraint to keep m < 1.0. The numeric value of efficiency when m = 1.0 is about 33 per cent. Thus AM has a maximum practical efficiency of about 33 per cent due to the limitations placed on modulation index.

At the receiving end, the information signal is recovered by selecting the upper side-band and downshifting it starting at zero. That is, in essence, what the receiver does when it demodulates.

Problems with Conventional AM

Conventional AM transmission has several problems:

- Bandwidth is wasted by having two identical side-bands on either side of the carrier.
- The efficiency is limited to 33 per cent to prevent distortion in the receiver when demodulating.
- The carrier signal is present even if nothing is being transmitted.

Single Side Band (SSB)

The AM described above is called double side-band AM or DSBAM. It is more efficient to transmit only one of the side-bands (called single-side-band), or even a part of one of the side-band. To make that happen, a band pass filter is added in the transmitter system to remove the lower side-band (LSB) and the carrier from the spectrum. The remainder is transmitted Fig. 5.11.

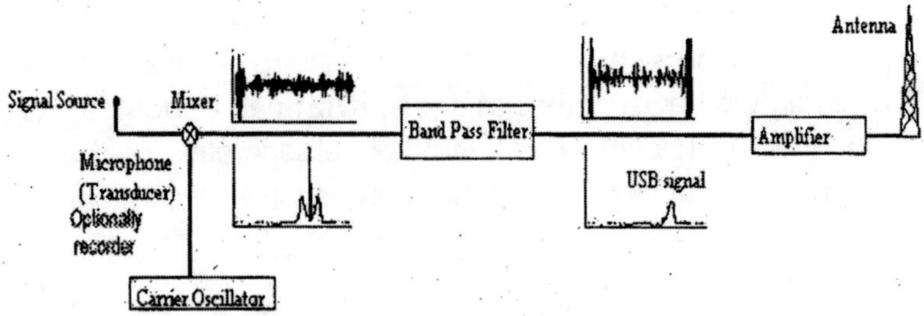

Fig. 5.11. Principle of SSB Transmission.

This technique is applied in commercial analog television, which is known as Vestigial Side-Band AM. The TV video signal has a bandwidth of about 4.25 MHz, but only 1 MHz of the lower side-band of the signal is transmitted. The 6 MHz per channel of TV signal has 0.75 MHz for the FM sound signal.

The Benefits of SSB

By eliminating the one side-band and carrier from transmission, the bandwidth gets reduced by half. The bandwidth in a SSB system is expressed by the formula:

$$BW = fm,$$

where, fm is the maximum modulating frequency.

In SSB, since the bandwidth of the transmitted signal is halved, it is possible to double the number of channels (or stations) in the same frequency band. If the carrier is also filtered out, there is no transmission unless information is sent. The efficiency is the power in the side-bands divided by the total power. In SSB, only side-band information is transmitted, accordingly the efficiency is 100 per cent in theory. However, due to losses in the circuitry and antenna, a typical efficiency of 80-95 per cent is achieved.

STEREO AUDIO

AM radio channels can be received on both stereo and mono receivers. This is done by generating a modulating signal by adding a DSB version (carrier at 38 KHz suppressed) of the output of the difference between the Left and Right channels added to the sum of the Left and Right channels unmodulated. The resulting modulating signal has a bandwidth of about 60 KHz. A mono receiver gets the sum signal whereas a stereo receiver separates out the difference as well and reconstitutes the Left and Right channel outputs.

SHORT WAVE AM TRANSMITTER

A short wave AM transmitter is shown in Fig. 5.12. The Control Panel of the SW transmitter is shown in Fig. 5.13. Output stage valve along with the cooling arrangement is shown in Fig. 5.14.

Fig. 5.12. A SW Transmitter (AM).

Fig. 5.13. Control Panel of SW Transmitter.

Fig. 5.14. Output Stage of the SW Transmitter.

FREQUENCY MODULATION (FM)

Armstrong proposed FM in 1933, which is also called angle modulation scheme. FM (Frequency Modulation) is a system of radio transmission in which the RF (Radio Frequency) carrier wave is modulated so that its frequency varies with the audio signal being transmitted. Initially there was much opposition by De Forest, but by 1949, FM was well established as a mode of high quality broadcast. FM was inspired by phase modulation but it has proved to be more useful partly for its ease of generation and decoding. The main advantages of FM over AM are:

- Improved signal to noise ratio (about 25 dB) with reference to the man-made interference.
- Smaller geographical interference between neighbouring stations

- Less radiated power.
- Well-defined service areas for given transmitter power.

In a frequency-modulated system, the frequency of the carrier is varied according to the modulating signal Fig. 5.15. For example, positive peaks would produce a higher frequency, while negative peaks would produce a lower frequency. At the receiving end, a limiting circuit removes all amplitude variations from the signal, and a discriminator circuit converts the frequency variations back to the original signal. In this way, the effects of amplitude noise are minimized. Since the recovered audio is dependent only on the frequency, and not on the strength, no compensation for different signal levels is required, as is the case with AM receivers.

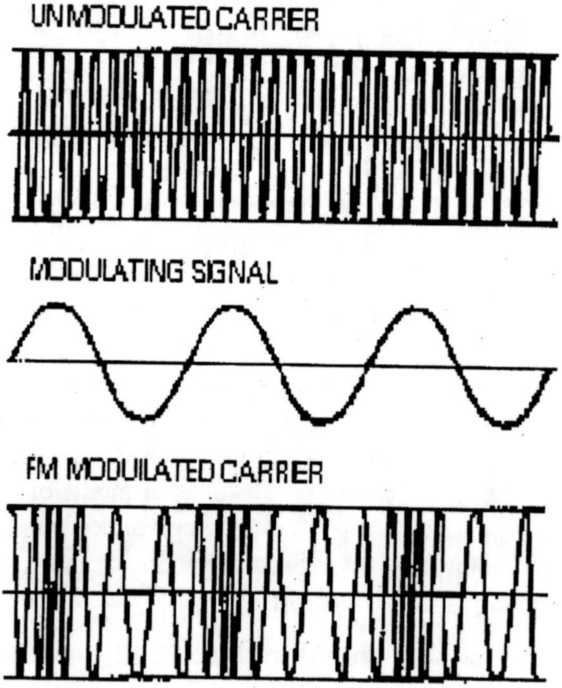

Fig. 5.15. FM Signal.

The signals in mathematical form are given below:

Information:

$$m\ (t) = \cos w_m t$$

FM: $A \cos(w_c t + \Delta w \int_0^t m(t)dt)$

Here the signal $m(t)$ is assumed to be normalized so that the maximum of the integral is 1.

$$\Delta\Omega$$

is called the frequency deviation of the modulation scheme. The index of modulation of an FM signal is expressed as below:

$$\beta := \frac{\Delta\omega}{\omega_m}$$

Fig. 5.16, 5.17, and 5.18 are examples of what FM signals look like in the time domain for a message signal of the form

$$m\ (t) = \cos \omega_m t$$

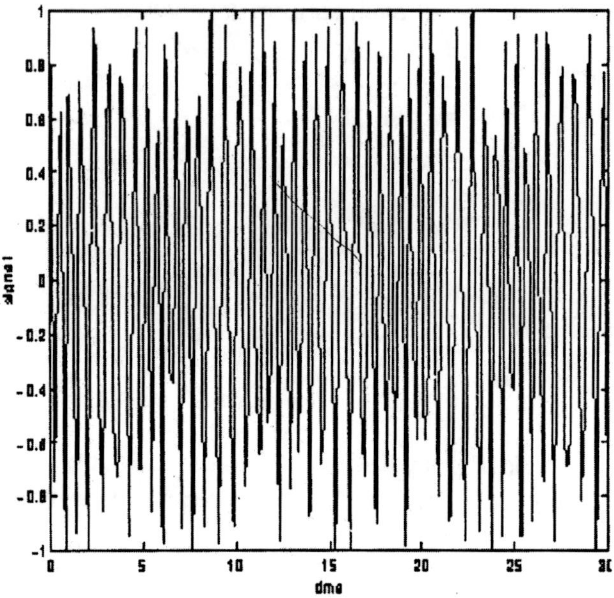

Fig. 5.16. FM modulation with modulating frequency 1, carrier frequency 10 and modulation index 2.

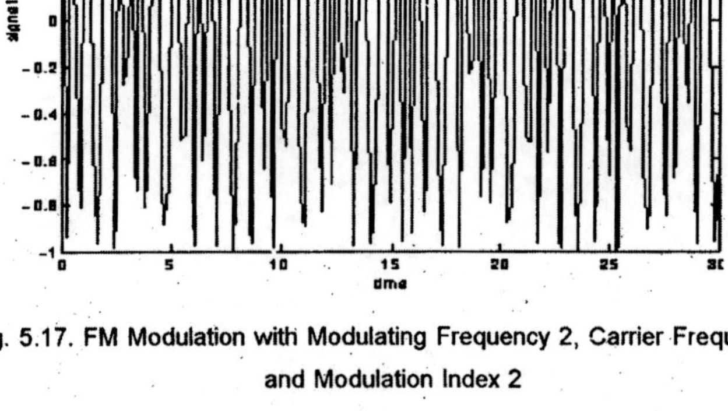

Fig. 5.17. FM Modulation with Modulating Frequency 2, Carrier Frequency 10 and Modulation Index 2

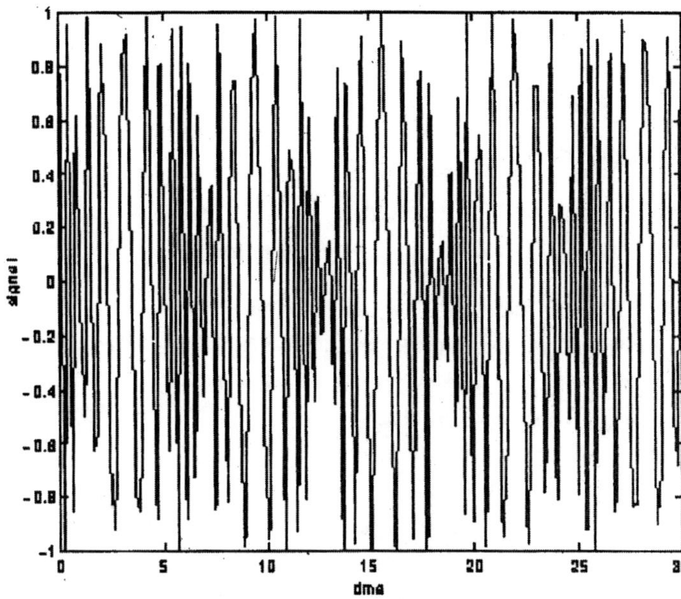

Fig. 5.18. FM modulation with Modulating Frequency 1, Carrier Frequency 10 and Modulation Index 5.

Bandwidth of an FM waveform

In general the determination of the frequency content of an FM waveform is complicated, but when β is small, for a sinusoidal signal m(t), the bandwidth of the FM signal is

$$2\omega_m$$

On the other hand when β is large, the bandwidth is determined (empirically) to be

$$2(\beta+1)\omega_m$$

Frequency range of FM

The allocated range for FM is from 88 MHz to 108 MHz. This range is in between the frequency ranges for the TV channels 1-6, and 7-12. For a typical radio station, β varies from about 150 at low frequencies like 50 Hz to 3.75 at high frequencies (20 KHz), and a very rough bandwidth figure is 200 KHz. Thus, a FM centre operating at 102.5 MHz goes from 102.4 MHz to 102.6 MHz.

Disadvantages of FM

- Much more bandwidth (as much as 20 times).
- More complicated receiver and transmitter.

A typical FM transmitter and its salient features

Technology has taken many strides and now a days, a single tube FM transmitter of 10 kW power are available Fig. 5.19[5].

Fig. 5.19. A Single Tube 10 kW FM Transmitter.

The performance features of the transmitters are completely self-contained broadband solid state IPAs, single high-efficiency PA tube, low noise, high-efficiency air-cooled system for longer tube life etc. The operational features include true proportional VSWR foldback, "soft-start" features, turn-on into high VSWR loads, emergency 250 watt RF patching, remote control interface, consistent RF output power to ±2 per cent, etc.

SUBCARRIERS: THE "HIDDEN SIGNALS"

Many FM broadcast stations now transmit subcarriers on their signals. Subcarriers are signals that are modulated on the carrier, just like the normal audio signals, except that they are too high in frequency to be heard. Normal audio signals range in frequency from 20 to 20,000 hertz. Most subcarriers start at 56,000 hertz (56 kHz). These subcarriers are themselves modulated, sometimes with audio signals, such as background music, or various forms of data viz. stock quotes, weather reports, news, sports and even paging signals. There is no limit to the variety of data that can be sent on a subcarrier signal, and broadcasters are finding new things to send all the time.

FM STEREO

In FM stereo system, the main audio, which is monophonic and stops at around 18 KHz. At 19 KHz is a pilot signal (a subcarrier) and then there is a special stereo subcarrier centered at 38 KHz. The receiver uses this special subcarrier in combination with the monoaural signal to produce the left and right channel audio signals. The 19 KHz pilot signal is used as a reference so that the receiver can properly recover the 38 KHz stereo subcarrier.

DIGITAL MODULATION SCHEMES

The digital signal arises from several sources, viz. encoding ASCII characters, sampling and quantizing speech or a video signal, etc.

In digital modulation, digital signal of 1s and 0s are prepared for transmission. The simplest digital modulation scheme is called On Off Keying (OOK) where a signal is turned on in response to a 1 and off in response to a 0. This scheme shown in Fig. 5.20 is a very popular scheme for transmitting signals in fiber optic cables.

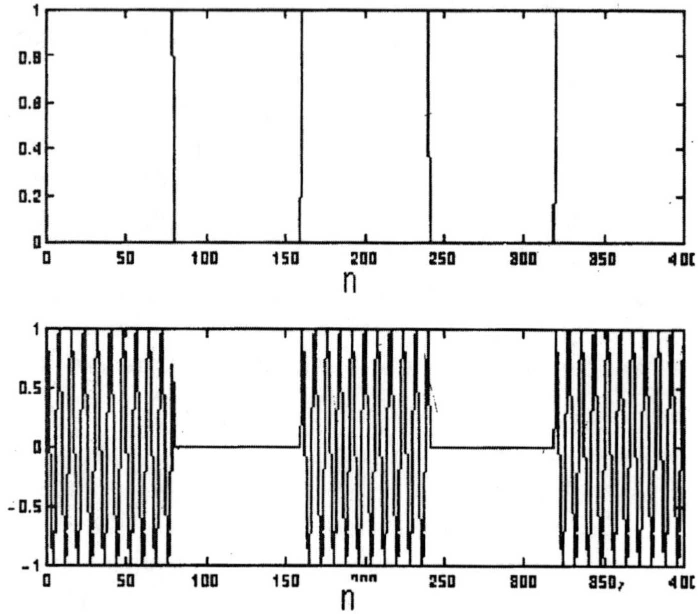

Fig. 5.20. OOK Modulation.

PULSE MODULATION SCHEMES

In pulse modulation, the carrier signal is used as a train of square pulses, or raised co-sine pulses, or sync function pulses (Nyquist). The choice of this pulse train is quite interesting from the standpoint of energy and spectral content consumption. Considering square pulse trains, the characteristics can be varied in terms of its amplitude or width or position of leading edge. The modulation schemes that vary amplitude and width are called Pulse Amplitude Modulation (PAM) and Pulse Width Modulation (PWM) respectively. The Fig. 5.21 below shows a passband PAM, in that the PAM pulse train is actually also multiplied by a carrier frequency sinusoid, so as to enable transmission at a higher frequency.

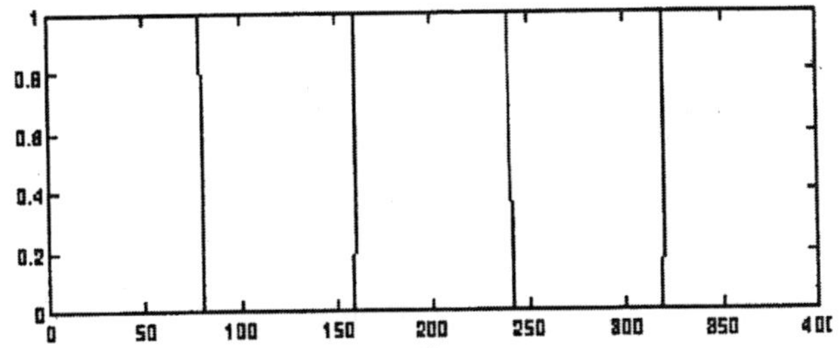

Fig. 5.21 (a) Message Signal with Bit Sequence 1010.

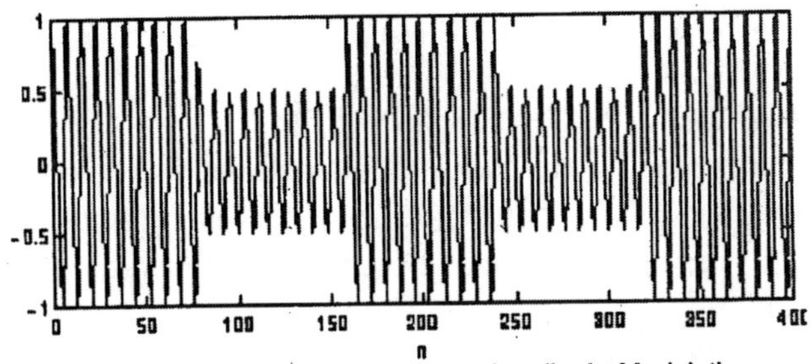

Fig. 5.21(b) Passband Pulse Amplitude Modulation.

This basic scheme can be made more sophisticated by using several amplitude levels: for example one can group the bits into groups of two, i.e. 00, 01, 10 and 11 and have four different amplitude levels for each of these groups. This is referred to as Quadrature Pulse Amplitude Modulation (QPAM or QAM for short). QAM actually is used for alphabets of size other than 4. Thus for example 2400 Baud full duplex modems use 16 QAM (corresponding to grouping 4 bits together). 9600 Baud, 14,400 Baud, 28,800 Baud modems use 32 QAM, 128 QAM and 1024 QAM respectively (along with trellis coding).

PWM is illustrated in Fig. 5.22 where the input information signal is the same as shown in Fig. 5.21 (a). The circuitry required to generate this is complicated, but it is extremely conceptually important since there is good reason to believe that neurons transmit information using PWM spike trains.

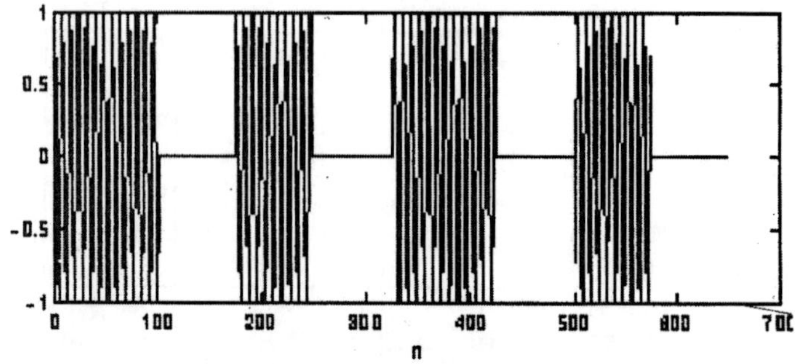

Fig. 5.22. Pulse Width Modulation (Longer pulses to 1, shorter pulses to 0).

A DIGITAL MODULATED TRANSMITTER

A Harris digital amplitude modulated medium wave transmitter is shown in Fig.5.23. Digital amplitude modulation samples incoming audio in a high-speed analog-to-digital (A/D) converter at a rate that is more than 10 times greater than that is used for compact discs. The output of the A/D converter is a 12 bit digital "word" that generates an amplitude-modulated signal by turning solid state RF power amplifier modules on (full voltage) or off (no voltage) to directly synthesize the RF envelope.

Fig. 5.23. Harris Digital Transmitter.

A 12 bit companded system of equally weighted and unweighted "steps" further varies amplitude. The six most significant bits control equally-weighted (equal voltage) solid state amplifier modules, and the six least significant bits control unequally-weighted (or binary) amplifier modules. Binary amplifiers, which are weighted at 1/2, 1/4, 1/8, 1/16, 1/32 and 1/64 voltage, can be used in combination with other binary amplifiers to create 63 different RF levels between each step. 6

KEYING SCHEMES

In the PAM system, the amplitude of the carrier system is varied or keyed in response to patterns of 1s and 0s. In a similar fashion, either the frequency or phase of a carrier signal can be keyed to obtain Frequency Shift Keying (FSK) and Phase Shift Keying (PSK). In essence Frequency Shift Keying (FSK) stands for the process of keying between two different frequencies Fig. 5.24.

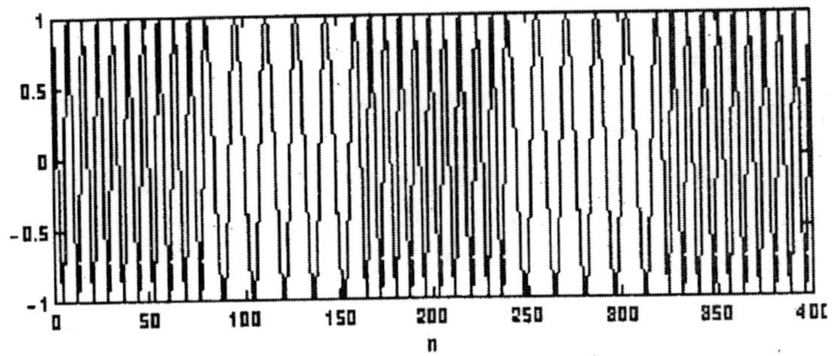

Fig. 5.24. Frequency Shift Keying (1 kHz to 1, 500 Hz to 0) with the Input Information Signal Shown in Fig. 5.21(a).

As in the case of PAM, we can group the bits and then do frequency shift keying and this is referred to as QFSK or 4 FSK, 8 FSK, etc. For example a full duplex 300 baud modems may use 2 FSK (1070 Hz for a 0 send, 1270 Hz for a 1 send; 2025 Hz for a 0 receive and 2225 Hz for a 1 receive).

PSK is a technique for switching phases in response to the signal (Fig. 5.25). As before there are quadrature versions of PSK which are referred to as either QPSK or 4 PSK, 8 PSK, etc. The

1200 Baud full duplex, 2400 Baud full duplex and 4800 baud full duplex modems use 4 PSK or 8 PSK.

Fig. 5.25. Phase Shift Keying (0 Pphase Shift to 1, 100 Phase Shift to 0), with the Input Information Signal is the Same as Shown in Fig. 5.21 (a).

In channels where there is rapid fading such as mobile radio (wireless) channels it is useful to use PSK modulation, since the decoder does not need to keep track of the intensity of the received signal. For greater noise immunity in these applications, a differential modulation scheme is used. In this scheme, one encodes the changes in the group of bits being encoded. Thus, for example digital cellular radio uses a differential PSK-4 (DPSK-4) scheme.

AMPLIFIER

Amplifier is a device that responds to a small input signal (voltage, current, or power) and delivers a larger output signal that contains the essential waveform features of the input signal. Amplifiers of various types are widely used in such electronic equipment as radio and television receivers, high-fidelity audio equipment, transmitters and computers. Amplifying action is provided by electromechanical devices (e.g., transformers and generators) and vacuum tubes, but most electronic systems now employ solid-state microcircuits as amplifiers. Such an integrated circuit consists of many thousands of transistors and related devices on a single tiny silicon chip.

An electronic amplifier can be designed to produce a magnified output signal identical in every respect to the input signal. This is a linear operation. If the output is altered in shape after passing through the amplifier, amplitude distortion exists. If the amplifier does not amplify equally at all frequencies, the result is called frequency distortion, or discrimination (as in emphasizing bass or treble sounds in music recordings).

A single amplifier is usually insufficient to raise the output to the desired level. In such cases the output of the first amplifier is fed into a second amplifier, whose output is fed to a third one, and so on, until the output level is satisfactory. The result is cascade, or multi-stage amplification. Long-distance telephone, radio, television, electronic control and measuring instrument, radar, and countless other devices all depend on this basic process of amplification. The overall amplification of a multi-stage amplifier is the product of the gains of the individual stages. When the power required from the output of the amplifier is so large as to preclude the use of electronic devices, magnetic amplifiers find wide application.

There are various schemes for the coupling of cascading electronic amplifiers, depending upon the nature of the signal involved in the amplification process. Solid-state microcircuits have generally proved more advantageous than vacuum-tube circuits for the direct coupling of successive amplifier stages. Transformers can be used for coupling, but they are bulky and expensive.

TV TRANSMITTERS

Television transmitters are classified in terms of operating band, power level, final-tube type, cooling method, etc. The visual transmitter refers to that portion of the television transmitter, which accepts the video signal, modulates a radio frequency (RF) carrier, and amplifies the signal to feed the antenna system. Present-day practice is to use an exciter operating at or near the intermediate frequency (IF) followed by an up-converter, which translates the IF signal to the desired channel frequency. It is also common practice to perform most of the various distortion corrections at IF, although some may be done at baseband and RF. The differences between low VHF, high VHF, and UHF, as far as the transmitter design is concerned, are mainly in the up-converter and the output power level. Low VHF are channels 2 to 6, and high VHF are channels 7 to 13. UHF channels used are 14 to 83, except for channels 70 to 83, presently assigned to mobile radio service. The aural transmitter is that portion of the television transmitter that accepts the audio signal, FM-modulates a RF carrier, and amplifies the signal to feed the antenna system.

TRANSMITTER CLASSIFICATION

Power Output. Output power refers to the peak power of the visual transmitter. *The power is the transmitter output power, less the feedline losses times the power gain of the antenna.* Low VHF power output sufficient to produce 100 kW, can be 35 kW with an antenna gain of 4, to 10 kW with an antenna gain of 12. High VHF power output can be 50 kW with an antenna gain of 8, to 30 kW with an antenna gain of 12, to produce maximum power. UHF output power, in order to achieve 5000 kW, could use a 220 kW transmitter with an antenna gain of 25. Most UHF stations operate considerably below the maximum. The transmission line losses vary according to frequency and length and will affect the final ERP. Power output and height limitations for transmitters in various bands are given in Table 5.1.

Modulation Level. Transmitters classified by level are of the IF type. Some of the technical reasons for using IF modulation is the ease of correcting distortions and ease in vestigial side-band shaping.

Table 5.1. Power and Height Limitations

Band	*Channels*	*ERP, kW*	*Height, ft(m)*	*Zone*
Low VHF	2-6	100	1000 (305)	1
			2000 (610)	2, 3
High VHF	7-13	316	1000 (305)	1
			2000 (610)	2, 3
UHF	14-69	5000	2000 (610)	1, 2, 3

ERP = Effective radiated power.

Source: Federal Aviation Administration.

Final-Stage Tube Type. Generally Tetrodes are used for VHF transmitters and for lower-level UHF transmitters below 5 kW. Klystrons are used for UHF above 10 kW. The choice is dictated by the tube that will best operate at the required frequency, taking into account the initial cost, the operating cost per hour, and the cost of the auxiliary equipment.

Final-Tube Cooling: Cooling can be by water, air, vapour, or a combination of these. Operating frequency, operating costs, tube life, noise, and efficiency determine the final selection.

Translators : Low-power stations, or translators, re-broadcast another station's programmes on another frequency, usually in the grade B service area of the mother station. They do not originate their own programs, and their power output is limited to 100 W VHF and 100 W UHF.

Low-Power Stations: These are, in effect, translators, except that they originate programmes. The power limitations are 100 W VHF and 1000 W UHF.

TRANSMITTER DESCRIPTION

Visual. The input signal is fed first to a video processing amplifier and equalizer unit. The processing amplifier performs the following functions:

- Video input amplification
- Receiver delay equalization
- Video clamping and hum cancellation
- Phase-delay compensation
- White clipping
- Sync regeneration

The processed video is converted to amplitude modulation of an IF frequency. The carrier is either generated by a crystal oscillator, or by a frequency synthesizer. The modulated IF is band-shaped in a vestigial side-band filter, typically a surface-acoustic-wave (SAW) filter. Envelope-delay correction is not required because of its inherent stable characteristics. Envelope-delay compensation may, however, be required for other parts of the transmitter. The modulated visual-IF signal is processed for distortion corrections and subsequent RF processing in the following stages.

- Power amplifier linearity correction
- Power amplifier group delay correction
- Driver amplifier linearity correction
- Incidental phase modulation correction
- RF up-conversion
- Channel bandpass filtering

The IF power amplifier and the final power amplifiers, thereafter, raise the power level to the desired requirement. A RF sample, obtained from a directional coupler installed at the output of the transmitter, is used in an automatic power-level control circuit located in the RF processor section. A colour-notch filter is used to provide the necessary additional attenuation of the lower side-band, and a harmonic filter is used to attenuate out-of-band. Additional sample couplers provide signals for the following monitoring facilities:

- VSWR metering
- VSWR protection
- RF for the demodulator
- RF for the spectrum analyzer and side-band units

AURAL

The audio signal is pre-emphasized and used to frequency-modulate the IF carrier generated in the oscillator section. The modulated signal is then up-converted to the final RF carrier frequency and passed to the intermediate power amplifier (IPA) and power amplifier (PA) stages. A harmonic filter is used to attenuate out-of-band radiation.

The visual and aural output signals are fed to a hybrid diplexer where the two signals are combined to feed the antenna. For those installations that require dual-antenna feedlines, a hybrid combiner with quadrature-phased outputs is used.

A typical RCA transmitter system is depicted in Fig. 5.26.

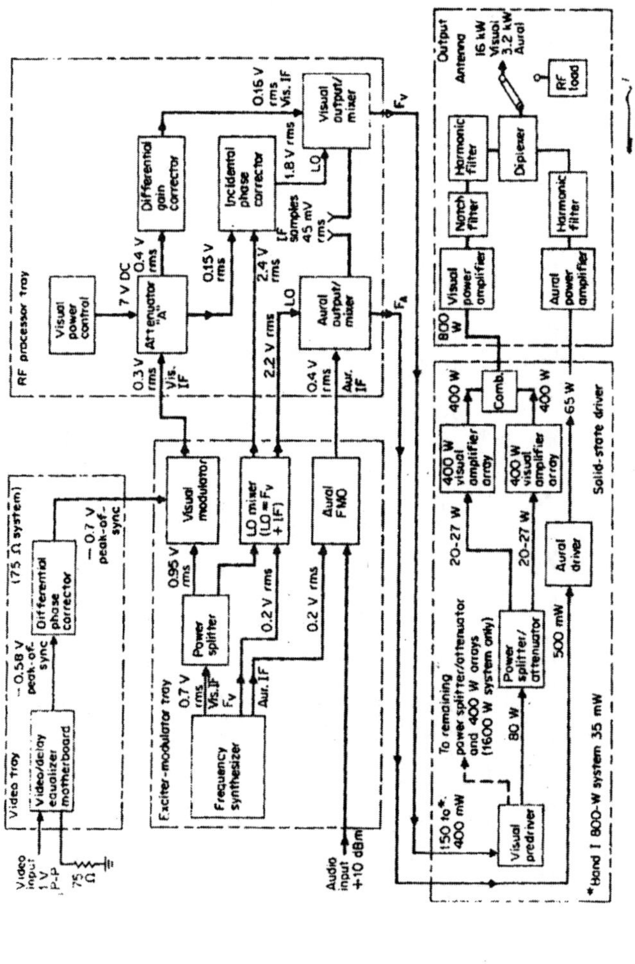

Fig. 5.26. A Typical RCA Transmitter System.

LATEST TRENDS IN TV TRANSMITTERS

High-power solid-state TV transmitters operating in the VHF band, continuously tunable from 170 MHz to 230 MHz, are available now-a-days. High-gain power transistors open up new solutions for power amplifiers and couplers, which are a cost-effective alternative to tube amplifiers in the 10 kW and 20 kW power class, too. The transmitters are available for colour TV broadcasting in PAL, PALplus, NTSC or SECAM colour systems as well as for dual-sound/stereo broadcasting according to the IRT or NICAM method. Microprocessor controlled components set new standards in ease of operation. A typical TV transmitter from Rohde and Schwarz is shown in Fig. 5.27.

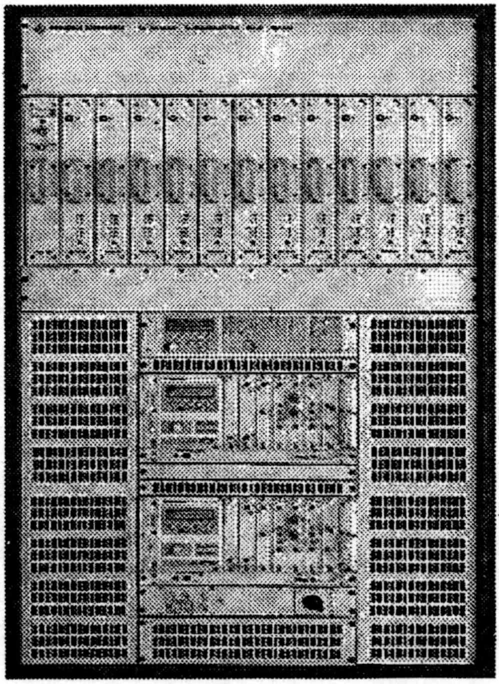

Fig. 5.27. A typical solid state TV transmitter from Rohde and Schwarz.

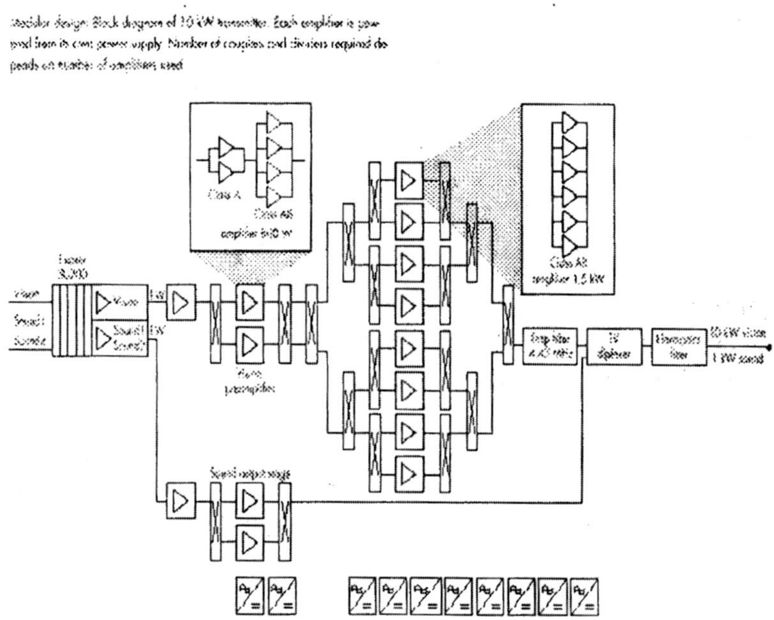

Fig. 5.28. Principle of operation of TV Transmitter.

PRINCIPLE OF OPERATION

Principle of operation of TV Transmitter is shown in Fig. 5.28. Two vision amplifiers are combined via a two-fold coupler to form a 2.5 kW power unit. The power units are cascaded in pairs to form a 5 kW or 10 kW unit. The RF powers of the two sound amplifiers are also added in a two-fold coupler. A TV diplexer combines the vision and sound power and takes it to the common antenna line. The colour subcarrier trap (-4.43 MHz from vision carrier) is connected ahead of the TV diplexer.

Exciter

The exciter produces from the video and audio signals, standard RF vision and sound signals. The exciter is designed for split amplification and for dual-sound transmission according to the IRT or NICAM 728 method. Comprehensive control and monitoring facilities protect the transmitter from overloads or failures even with non-standard input signals and operating states. The RF levels at the transmitter output are kept constant within a wide range with the aid of sync pulse regeneration and amplitude control.

Amplifiers

Drivers

The single-stage driver amplifiers are fed from the exciter (maximum. 1 W). The transistors of the drivers are configured for class A operation. The maximum output power of each driver is 10 W. A monitoring circuit signals any failure of the driver stage to the exciter.

Pre-amplifier

The pre-amplifier is used for vision and sound amplification. In the case of sound amplification the pre-amplifier is the output stage. Maximum. output power is 800 W (sync) or 700 W (cw).

The output of the drivers (maximum 10 W) feeds the pre-amplifiers. The pre-amplifiers have two cascaded stages. The first stage is made up of two transistors in class A, whereas the second stage has four transistors in class AB. RF power splitting and combining is implemented by means of Wilkinson circuits in stripline technique. RF testpoints are provided at the pre-amplifier outputs for monitoring reflections and measuring the forward power. With an analog multiplexer, which is controlled from a monitoring facility, the basic parameters of each transistor, i.e. current and supply voltage, can be measured during operation. In case of over-temperature or reflection, the power supply is switched off to protect its amplifier.

Output Amplifier

Eight identical amplifier modules implement the output vision amplifier. Each stage has six RF transistors operating in class AB, which are driven by a system of Wilkinson bridges. The amplifiers like the pre-amplifiers are equipped with monitoring facilities for over-temperature and reflection and RF testpoints for parameter measurement. Maximum. output power is 1500 W (sync).

Control and Monitoring

The microprocessor based transmitter control unit is the core of the transmitter. It provides for a correct switch-on sequence of the cooling system, power supply and exciter power and monitors the air flow, the inlet and outlet temperature as well as the operating status of amplifiers, power supply units and of the entire transmitter. During active operation, the RF output and reflected power for vision and sound, the power absorbed by the diplexer as well as the inlet and outlet air temperature can be called up for display. The display field comprises a separate readout for checking menu-selected operating parameters. The user can thus monitor typical parameters such as the transmitter output power for vision, sound 1 or 2 or the outlet air temperature. Any fault occurring in the transmitter is immediately displayed with top priority. Up to 40 transmitter faults are stored with time and date.

All relevant messages and commands are available at the parallel remote-control interface. A bitbus (864-1 IEC) is optionally available as a serial interface for commands and messages.

TRANSMITTER PERFORMANCE CHARACTERISTICS

A variety of degradation occurs in the received picture and sound, resulting from reflections and terrain limitations. These are in terms of poor resolution, colour fringing, hue shift, luminance shift, and poor audio quality. Among the critical performance parameters causing these types of degradations are incidental phase modulation, differential phase and gain, frequency response, envelope delay, quadrature distortion, luminance-to-chrominance shift, audio response, and distortion.

There are two broad classifications of signal distortions in television systems, linear and nonlinear. Linear distortions occur which are independent of picture level, while nonlinear distortions vary with the amplitude of the picture signal.

Linear distortions usually occur in systems with incorrect frequency response and are divided into four general categories:

Short-time distortions affecting horizontal sharpness

- Line-time distortion causing horizontal streaking
- Field-time distortion causing errors in the low-frequency response and vertical shading in the picture
- Long-time distortions in the direct current to extremely low-frequency range causing flicker and slow changes in brightness of the picture.

Incidental phase-modulation distortion results in noise in the audio reproduction and colour distortion in the picture. It occurs when the quadrature side-band components of the vestigial side-band television system are unequal, causing the instantaneous resultant of the two sideband vectors around the carrier change in phase, and appearing as though the picture and colour subcarriers are being phase-modulated. In other words, the normal AM picture and colour subcarriers appear to be FM signals.

Nonlinear distortions occur with changes in the average picture level (APL). Luminance nonlinear distortion occurs when the gain of the system varies as the signal changes from black to white. Differential gain distortion is the change in the amplitude of the chrominance signal as the amplitude of the luminance changes from black to white. Differential phase distortion is the change in the colour subcarrier phase as the luminance signal changes in amplitude.

IF Group Delay : Group delay is the distortion that results when a group of related frequencies differ in arrival time when passing through a circuit. Some examples of this are distortion due to the restrictions in frequency response of the vestigial side-band filter, the aural sound notch, the colour-notch filter, and the average receiver response.

Low-Frequency Delay: The characteristic LF delay has a significant effect upon the luminance signal, while high-frequency delay primarily affects the chrominance signal.

Quadrature Distortion: This is the apparent phase shift of the visual carrier when energy exists unequally in the two side-bands.

Differential Phase and Incidental Phase Modulation Distortion: Differential phase is the change in phase of the colour subcarrier for a change in the brightness level. Incidental phase modulation is the undesirable phase modulation of the visual carrier and its colour subcarrier is relative to the visual carrier being amplitude-modulated.

Differential Gain: Differential gain distortion occurs when the amplitude of the chrominance signal changes with respect to the amplitude of the luminance signal. This causes a change in hue saturation as the luminance varies between black level and white level.

Audio Response and Distortion: In new transmitters, amplitude-frequency response deviation and distortion are less than one per cent.

NEW TRANSMISSION SYSTEM

Transmission system in the broadcast scenario deals with the distribution of signal via terrestrial, satellite or cable. Till early 90s, the main delivery of broadcast signals was through terrestrial analog transmitters. The satellites were used for linking of terrestrial transmitters. The development was mostly towards increase in efficiency and signal quality as compared to the physical links. The recent trend is towards digital broadcasting using terrestrial modes or direct satellite broadcasting for the direct to home (DTH) service. In the following sections, few of the latest developments taking in this area have been described.

Digital Audio Broadcasting (DAB)

The prospect of sending higher quality audio to portable receivers is spurring development of digital audio broadcasting (DAB). Europe has led the way with its Eureka 147 project[7] The project was launched in 1986 and continued up to 1994. The main emphasis of the project was to carry out DAB transmissions test and also consider the new communication services as supplementary to the broadcast services (traffic information and management system, data transmission to specific group of users etc.) The DAB was carried out using an

OFDM (Orthogonal Frequency Division Multiplexing) packet in which six stereo programmes were transmitted with other data (Fig. 5.29). The project led to the detailed and comprehensive system definition and specification of the standards[8] The BBC and other European broadcasters have adopted the technical format. Japan started DAB satellite services in 1992 for small fixed panel antennas. BBC started its introductory DAB service in September 1995 in the frequency range 217.5 - 230 MHz. All India Radio conducted experiments with DAB in Delhi in 1995. A 125 watt transmitter was installed and the field trials were made. Broadcasters are investigating technologies to enable digital transmissions to share the same spectrum with existing FM and AM broadcasts. The advantages to the listener of the DAB system over FM are very impressive. The signal is very robust and spectrally efficient, virtually immune to multipath and fading effects, and offers nearly CD quality like sound. Service flexibility is another advantage of DAB. The multiplex can be dynamically configured to provide a wide range of bit-rates—more services at lower quality or fewer at high quality. Data such as lyrics and phone-in numbers can also be transmitted with the audio. Programme labeling, graphics and traffic messages are among the other enhanced features available. The data broadcasting capability also opens the way for a range of new services and commercial opportunities such as links to computers, visual displays for in-car navigation systems, or even Internet-style pages. NTL has recently broadcast such data at a recent radio festival in Birmingham, UK. The company also demonstrated an innovative triple simulcast on AM, FM and DAB to demonstrate the different subjective audio quality available from each transmission system. Using custom-programmed editorial, a programme stream was simulcast on 1323 KHz AM., 105.4 MHz FM and DAB Channel 11D (222.064 MHz). The DAB multiplex was filled by identical programming coded at a variety of rates from 256 to 32 kilobits per second, so that the direct comparisons could be made between conventional analogue and the flexibility of DAB. A block diagram of a DAB receiver is given in Fig. 5.30. DAB is much more spectrum efficient since a single frequency can be shared for the whole of the national terrestrial network.

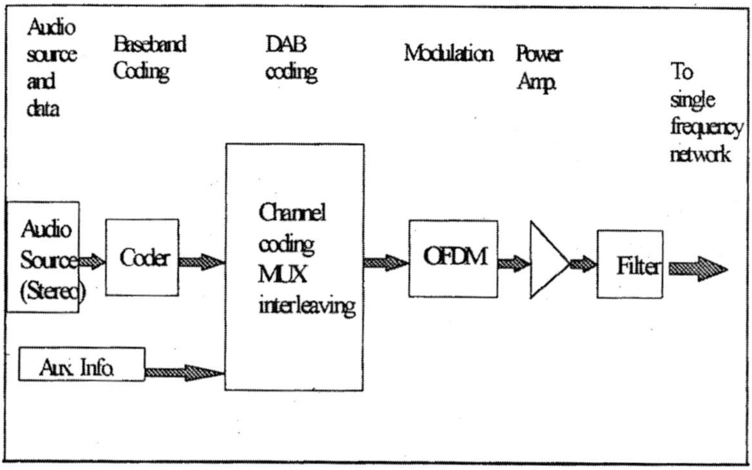

Fig. 5.29. Block Diagram of a DAB transmission System for mobile reception.

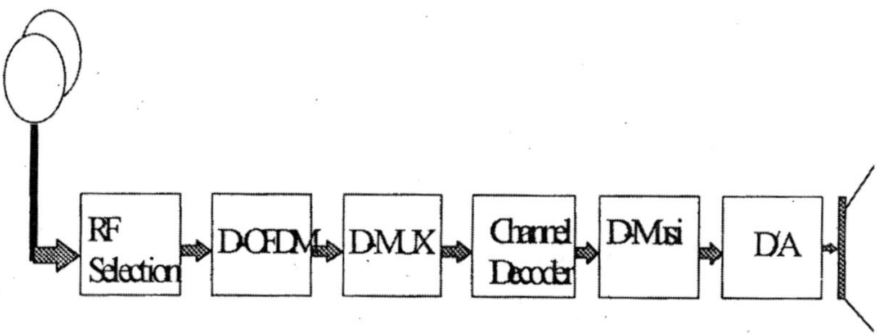

Fig. 5.30. Block Diagram of a DAB Receiver

To sum up, DAB provides multiplexes or 'packages' of up to six stereo services at a time in a rugged digital format which delivers consistent high-quality reception on-the-move and in fixed locations with very modest receiving antennas.

High Definition Television

Around the world, High Definition Television (HDTV) systems are being brought to the fore. HDTV is a television format with a new screen aspect ratio of 16:9 (the current is 4:3) and capable of reproducing much greater detail (5 to 6 times more) and colour information (ten times) than existing broadcast systems. When projected on a 16/9-foot screen and observed from a normal viewing distance, the picture detail in HDTV/DTV appears to equal what is normally attained by projected 35 mm motion picture film. HDTV should not be confused with wide screen variants of PAL (PALplus), NTSC or SECAM where the screen shape may change but the quality improvement is small compared to HDTV.

The standard for analog HDTV was defined in 1970 with 1125 lines scanning rate, line interlacing, 5:3 aspect ratio and 60 Hz field frequency. The trials were conducted in Japan using BS-2B satellite in 1989[9] and the experimental transmission started in 1991. Japan has begun limited satellite transmission but high receiver costs combined with lack of HDTV programmes are resulting in limited set penetration. There is no agreement for the world HDTV standard. In Europe 1250/50, with its simple relationship to 625/50, is favoured while in the USA, 1050/59.94, with its relationship to 525/59.94, has been adopted. In Europe and Japan, the emphasis is on analog satellite transmission. In the USA, the Japanese standard with 1125 lines /60 Hz has been accepted. However, there is the problem of terrestrial TV channels having a bandwidth of 6 MHz and priority being given to terrestrial programme distribution over other transmission media. Analogue HDTV certainly cannot be implemented in the 6 MHz channel and therefore U.S. is focusing on digital terrestrial transmission systems using video compression. The only consensus so far is that transmission, for home viewing and contribution, will be digital and compressed, using MPEG-2. The USA has chosen its Grand Alliance system for advanced television. ITU-R has two production standards based on 1125/60 and 1250/50 formats[10].

Similar to the quality of sound gained from the compact disc (CD) HDTV sets a standard, in which high quality sound is available. The digital format selected for audio for DTV/HDTV is called 5.1 channel sounds and is based on the dolby digital surround-sound format. The 5.1 system consists of six discrete channels of audio: left, centre and right channels in front of listeners, and left-surround and right-surround at the sides. The 6th channel is a bass channel of limited frequency response (3-120 Hz). Although it is capable of producing a room-rattling bass, it only takes up one-tenth of a full-range audio channel. Hence, the system is referred to as 5.1.

By using 12:1 audio signal compression all six channels can be transmitted in a comparatively limited bandwidth. Even so, accommodating the full capabilities of 5.1 audio in TV production will require audio boards and mixers capable of six channels of audio. Although DVD players are capable of 5.1 audio, most homes are not equipped for this level of sound sophistication.

Fig.5.31 shows the transmission chain for HDTV broadcasting. The high quality video input is taken via a signal-processing unit before applying to the source encoder. The output data rate of 20 to 30 Mbits/sec is subjected to forward error correction (FEC) channel coding. In the subsequent multiplexer, the audio, sink data and control signals are added to the coded picture signal, whereupon a gross data rate of 30 to 40 Mbits/sec is obtained. The channel coded data are modulated onto the IF carrier and packed into a channel bandwidth of 7 to 8 MHz (6 MHz in US). A mixer up-converts to the RF carrier frequency, e.g. in Band I to Band V (47 MHz to 862 MHz) before the signals are applied to the terrestrial transmitter and antenna.

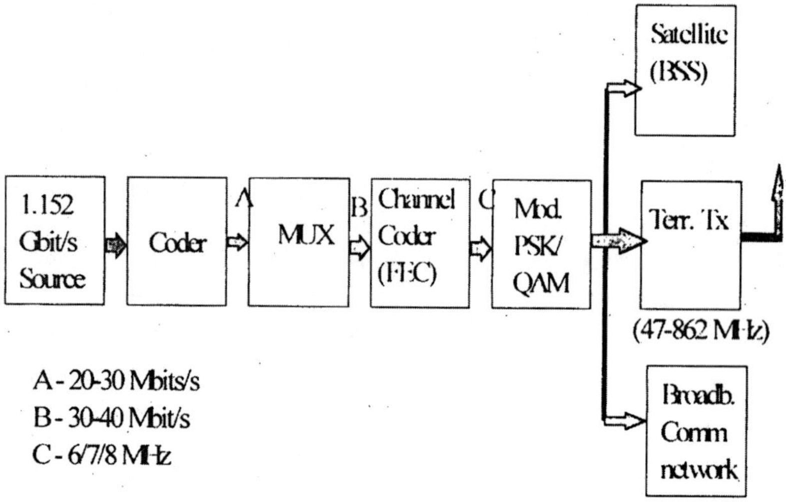

Fig. 5.31. Transmission Chain for Digital HDTV.

DIRECT BROADCAST SERVICE (DBS)

In the 1970s and 1980s satellite became the chief means of long distance radiocommunication and facilitated worldwide TV programme distribution in real time. In rural areas not serviced by terrestrial TV broadcast stations and cable TV systems, satellite signals made possible direct reception of TV from satellites by TVRO (TV Receive Only) receivers equipped with parabolic antennas with diameters between 2 and 5 meters. For almost two decades, TV programme delivery by satellite was done in the 4/6 GHz and 11/12.4 GHz bands. Doordarshan (TV India) is using this technique to provide centrally produced programmes to Low Power TV (LPTV) stations.

Development in advanced radio communications technologies and the offer of improved radio based services paved the way for direct TV and audio broadcasts from satellites (also known as Direct To Home—DTH service). Today, three technologies have been aggressively developed—the direct broadcast service (DBS), BSS-HDTV and BSS-Sound.

Direct broadcast satellite (DBS) was introduced in U.S. in late 1993 for the delivery of conventional television programming directly to the consumer, using frequency of 12.2-12.7 GHz for the BSS. The DBS system consists of a small 18 inch satellite dish, a digital integrated receiver/decoder (IRD) which separates each channel, decompresses and translates the digital signal, and a remote control. The dish never has to track the satellite so there is no waiting for the picture to come in[11] DBS programming is distributed in USA by three high power HS-601 satellites (DBS-1, DBS-2 and DBS-3), co-located in geo-synchronous orbit, 22,300 miles above the earth at 101° west longitude. DBS-1 features 16 number., 120 watt Ku-band transponders with DBS-2 and DBS-3 each configured to provide 8 transponders at 240 watts. DBS-1 delivers upto 60 channels of programming and approximately 20 channels of programming is provided from USSB (United States Satellite Broadcasting). DBS-2 and DBS-3 are used exclusively to provide approximately 175 channels. Each one of the transponders on the DBS-1 satellite can send more than 23 Mbit/s. The DBS-2 and DBS-3 satellites are even faster, at around 30 Mbit/s each.

DBS employs MPEG-2 technology, the emerging world standard for digital broadcasts. DBS is fully digital and "forward compatible" so that consumers can take advantage of emerging technologies, such as interactive services, 16 x 9 wide screen and HDTV (digital) broadcasts.

Programming comes from various content providers (CNN, ESPN, etc.) via satellite, fiber optic cable and/or special digital tape. Most satellite (analog communication satellites used for programme delivery) delivered programming is immediately digitized, and uplinked to the orbiting satellite. The DBS satellites re-transmit the signal back down to every earth station, or in other words, every little DSS receiver dish at subscribers' homes and business.

Broadcasting Satellite Service-HDTV is meant for HDTV but due to the problem of large bandwidth required for analogue HDTV, the service has not become operational. WARC-92 (World Administrative Radio Conference-1992) have allocated two frequencies: 17.3-17.8 GHz for Region 2 (Europe), and 21.4.22.0 GHz for Regions 1 (America) and 3 (Asia) for this service. These new allocations will become effective from April 1, 2007[12].

Broadcasting Satellite Service-Sound (BSS-Sound) is for high quality audio programming: BSS-Sound generally refers to the delivery of music, sports, news, etc. directly to consumers' radio via satellite. WARC-92 adopted three different allocations for BSS-Sound: 1452-1492 MHz, 2310-2360 MHz and 2535-2655 MHz. The service shall be received using portable radios. Such services will not be compatible with existing analog AM and FM radios and require consumers to purchase new radios to enjoy this new broadcast service.

THE GRAND ALLIANCE—ALL DIGITAL HDTV

In 1987, the FCC issued a ruling indicating that the HDTV standards to be issued that would be compatible with existing NTSC service, and would be confined to the existing VHF and UHF frequency bands. By the end of 1988, the FCC had received 23 different proposals for HDTV or EDTV standards. These were all analog (or mixed analog/digital systems like MUSE) and explored a variety of different options for resolution, interlace and bandwidth.

In 1990, the FCC announced that HDTV would be simultaneously broadcast and that its preference would be for a full HDTV standard. On May 31, 1990 General Instrument Corp. submitted the first proposal for an all-digital HDTV system. By December 1990, ATRC announced its digital entry, followed quickly by Zenith, AT&T, and MIT. Thus there were four serious contenders for digital HDTV, as well as a modified "narrow" MUSE and an EDTV proposal. During the following year, these systems were tested.

Subsequent to this, a recommendation was made to form a "Grand Alliance" composed of AT&T, GI, MIT, Philips, Sarnoff,

Thomson and Zenith. This Grand Alliance was formed with the intention to take the best features of the four systems and develop them into HDTV standard.

THE BASIC STANDARD

The Grand Alliance standard differs from all existing TV standards in three major ways. First, all digital standards are to be broadcast with a packet transmission. Second, it supports multiple formats. Third, it is designed to be primarily compatible with computers rather than existing NTSC televisions.

COMPRESSION

The compression algorithms use both a motion compensated and discrete co-sine transform (DCT) algorithm. The motion compensation exploits temporal redundancy. The DCT exploits spatial redundancy. MPEG-2 syntax was used for computer and multimedia compatibility. Dolby AC-3 digital audio compression was chosen with full surround sound.

The core of the Grand Alliance concept is a switched packet system. Each packet contains a 4 byte header, and a 184 byte data word. Each packet contains video, audio, or auxiliary information. For synchronization, the program clock reference in the transport stream contains a common time base. For lip sync between audio and video, the streams carry presentation time stamps that instruct the decoder when the information occurs relative to the program clock.

The terrestrial transmission system is 8 level vestigial side-band (VSB) technique. The 8 level signal is derived from a 4 level AM VSB and then trellis coding is used to turn the 4 level signals into 8 level signals. Additionally, the input data is modified by a pseudo-random scrambling sequence that flattens the overall spectrum. Cable transmission is by a 16 level VSB technique without trellis coding.

Finally, a small pilot carrier is added. This pilot carrier is placed so as to minimize interference with existing NSTC service.

The Grand Alliance system is clearly designed with future

computer and multimedia applications in mind. The use of MPEG-2 will permit HDTV to interact with computer multimedia applications directly. For example, HDTV could be recorded on a multimedia computer, and CD-ROM applications could be played on HDTV systems.

RADIO PROPOGATION

Medium wave signals between 0.5 and 1.7 MHz travel using ground wave during the daytime and skywave during the nighttime, especially during the hours of darkness. These signals begin to get very weak and noisy after 1000 km or so, during nighttime.

FM radio operates in many countries between 88-108 MHz. But transmitters using this frequency range only have a service area of up to 100 km. The signals travel in a straight line between the transmitting tower and the receiver's antenna.

The short wave spectrum, 1.7 - 30 MHz, lies between medium wave and FM. Under the right conditions signals on these frequencies can travel thousands of kilometers. This is due to a refractive series of layers in the earth's atmosphere called the ionosphere. Signals at the transmitter site are beamed up to the sky at a low angle. They are then refracted back to earth by the ionosphere. The signal then "bounces" off the earth's surface again, and after a succession of "hops" reaches the antenna of the listener's receiver. The refractive properties of the ionosphere are not perfect, which is why signals sometimes fade in and out.

The electrically conducting layers in the ionosphere interact with electromagnetic radio waves and bend them in much the same manner that glass or water bends light. The refractive index of a layer in the ionosphere depends on its electron density and has an effect on waves according to their wavelength. Layers with higher electron densities (higher altitudes where recombination is slower) have lower refractive indices. As a wave passes up from below into such layers they get refracted, or bent back down towards the lower electron density layers. The amount of bending depends on the incident angle and the frequency of the wave. Higher frequencies are less affected. Finally, in passing through conducting layers, energy in the wave vibrates the electrons thus transferring some energy into the layer thus being partially absorbed in the process.

For much of the radio spectrum the frequencies are simply too high to be refracted back to the earth given the range electron densities which exist and propagation angles required. This is not true in the HF band of frequencies between a few hundred hertz and about 30 megahertz. Depending on the natural variations in density of the ionosphere, radio waves either get refracted back to the earth or passed out into space. For the case of vertical incidence and a given ionospheric density there is a critical frequency above which no reflection is received. Inversely, for a given frequency there is a "critical" density needed to return the waves. The relationship is expressed by the following formula:

$$Ncc = 1.2 \times 10^4 \, v^2 \cos^2 a$$

where, Nec = electron density cm^2

v= frequency, and

a= incident angle measured from vertical

This expression is for the critical situation only. Where the angle is approximately known based on the circuit geometry and the frequency is known precisely, the density calculated will be the lowest density, below which, no return (reception) will be possible. A radio wave of a given frequency therefore will continue to travel upwards into an ionosphere until it reaches the elevation where the density falls to the "critical" level (Nec) where the wave will be turned back to earth.

For most HF radio communication the main reflecting layer is the F layer situated between 200 to 300 km high. The D and E layers are generally too weakly ionized to be of much importance except during the day over shorter distances. Ionization, although falling, persists through much of the night in the F layer as well. Intense ionization of the F layer causes it to warm and split a second, higher level during the day, returning to a single layer at night. There are also seasonal variations on the theme. The E layer, about 100-150 km up, can persist into the night to some extent and in addition there are occasional "clouds" or masses of E level ionization that can drift causing sporadic radio openings. The D layer at about 80 to 90 km is only useful for propagation during the day re-combining quickly after sunset. In the negative sense, it absorbs signals passing through it to or from the F layer.

ANTENNA

The purpose of antenna systems is to interface with the transmission medium (air) and transmit and receive electromagnetic energy. Antennas represent the first and last component in a wireless system. An antenna is a conductor of electromagnetic energy. It collects EM energy while receiving and radiates it while transmitting. Antennas are considered reciprocal devices, since an antenna can receive and transmit radio frequency energy. Antennas are passive devices, which have loss rather than gain. The antenna gain is achieved by concentrating energy in a particular direction and losing energy in another direction. An antenna appears to be a source of energy as a resonant circuit. Energy stored within an antenna's resonant circuitry is said to be part of an induction field. Radiation fields are energy transfer fed to or from the antenna through free space characterized by near-field energy (surrounding the antenna), and far-field energy (at some distance from the antenna). An antenna's far-field energy is usually portrayed as a plane wave at some minimum distance (approximately 10 wavelengths of the operating frequency) from the antenna.

OPERATION OF AN ANTENNA

When a transmission line is terminated in an open circuit it introduces an abrupt discontinuity to the incident voltage wave in the form of a phase reversal. The phase reversal results in some of the incident voltage to be radiated and not reflected back to the source. The radiated energy propagates away from the antenna in the form of traverse electromagnetic waves. The radiation efficiency of an open transmission line is extremely low. Radiation efficiency is the ratio of radiated power to reflected power. One way to achieve more radiation is to spread the conductors apart (dipole antenna). When the conductors are spread out in a straight line, the length determines the type of antenna. If the line is one-quarter wavelength, it is called a quarter wave dipole, or vertical dipole. An antenna radiating element when connected to a source produces an E field component, which is aligned to the radiating element. Also generated is a magnetic field (H field), this H field is concentric around the element's axis and perpendicular to the E field.

ANTENNA CHARACTERISTICS

With the use of Maxwell equations for current flow, the magnetic and electric fields generated by the antenna can be determined. Antennas are also characterized by a number of performance parameters, viz. Bandwidth, radiation impedance, beamwidth, etc. These parameters are briefly described below:

Bandwidth

An antenna's bandwidth refers to the usable frequency range over which the gain is within 3 dB of a nominal value.

Radiation Impedance

A lumped circuit element description of an antenna that relays the power through an antenna to some reactance by an equivalent current and voltage. The formula for determining radiating efficiency is given by

$$E = P_r/P_t.$$

Where, P_r = the antenna radiated power and

P_t = the total power supplied to the antenna.

Beamwidth

Beamwidth is the angular separation between the two-half power points (-3 dB) on the major lobe of an antenna's plane radiation pattern, usually taken in one of the principal planes. An antenna's main lobe radiation pattern can be thought of as the energy travelling in the primary direction of propagation.

Radiation pattern

A polar diagram representing field strengths at various angular positions is relative to an antenna. The front to back ratio is the front lobe power to the back lobe power. The line bisecting the major lobe or pointing from the centre of the antenna in the direction of maximum radiation is called the line of shoot.

Directive gain and power gain

Directive gain is the ratio of the power density radiated in a particular direction to the power density radiated to the same point by a reference antenna, assuming both antennas are radiating the same power. Directivity is the maximum directive gain.

$$D = P/P_{ref}$$

Where, D = directive gain,
P = power density at some point
P_{ref} = power density at same point with a reference antenna.

Power gain

Power gain is the same as directive gain except that the total power fed to the antenna is used, taking into account antenna efficiency. The power radiated from an antenna can never exceed the input power. An antenna simply concentrates its radiated power in a particular direction.

Effective Isotropic Radiated Power (EIRP)

EIRP is the equivalent power that an isotropic antenna would have to radiate to achieve the same power density in the chosen direction at a given point as another antenna. EIRP is expressed by the formula,

$$EIRP = P_r A_t \text{ watts}$$

Where, P_r = total radiated power
A_t = transmit antenna directive gain

or $EIRP(dB_m) = 10 \log(P_r/.001) + 10 \log(A_t)$.

or $EIRP = P_{in} A_p$

where, P_{in} = input power, and
A_p = power gain.

Antenna polarization

The polarization of an antenna is the orientation of the electric field radiated from it. Polarization is the sense of the antenna's E field motion relative to the earth. Vertical polarization is when the E field is perpendicular to the earth. Horizontal polarization is when the E field is parallel to the surface of the earth.

ANTENNA TYPES

Half wave dipole

Half wave dipole is also called a hertz antenna. It is a multiple of quarter wavelengths long and open circuited at the far end. This antenna is resonant. Standing waves exist along this antenna. To the standing waves the poles of the antenna looks like an open quarter section of transmission line. Therefore there is maximum voltage and minimum current at the ends, and maximum current and minimum voltage in the middle. If the antenna is ungrounded, the ground reflected wave appears to be radiating from an image antenna below the earth's surface. The ground reflected wave is 180° out of phase with the incident wave, and travels farther to reach the same point.

Grounded antenna

A monopole antenna one-quarter wavelength long,mounted vertically with the lower end connected directly to ground is called a Marconi antenna. This type of antenna consists of a quarter wavelength conductor above the earth, and the earth acting as the other quarter wavelength element. Electrically, the Marconi antenna acts like a hertz antenna.

Antenna arrays

Two or more antenna elements combined to form an antenna array are used to obtain good directivity and directional gain. Depending on the method used to excite the additional elements, antenna arrays are classified as parasitic, driven or phased. Driven elements are directly connected to the transmission line, and are driven by the source. On the other hand parasitic elements are isolated from the source and receive energy through mutual induction. A parasitic element that is longer than the driven element from which it receives energy is called a reflector. A reflector reduces the signal strength in its direction, and increases it in the opposite direction.

Parasitic arrays

In parasitic arrays, the secondary element is excited by mutual induction. When a parasitic element is placed one-quarter wavelength from the driven element, the radiated wave travels 90° to the parasitic element. It is reversed by 180° in the parasitic element, then travels 90° back to the driven element, and arrives back at the driven element in phase with the driven element. This produces a 5dB gain in the forward and backward directions at the expense of the side radiation. If only forward gain is required a director element is used. A director element is shorter than the driven element by a half wavelength. This makes the director element resonant at a higher frequency than the operating frequency. If backward gain is desired the use of a reflector is used. A reflector element is cut longer than the driven element by half wavelength. Driven arrays use no parasitic elements, the added elements are fed from the source.

Collinear array (A four element driven array)

In a collinear array, the elements are placed end to end. The transmission line is connected to each element, and, since the line length to each element is equal, the current in each element is in phase. When vertically oriented, these are used for omni-directional coverage.

Phased array

A phased array is antenna with two or more driven elements. The elements are fed with a certain relative phase, and they are spaced at certain distance, resulting in gain in some directions, and little or no gain in other directions.

ANTENNA LOCATION

For proper wireless performance the polarization alignment of the transmitting and receiving antennas must be the same. Polarization, free space loss, and propagation in solids affect radiation from an antenna. Free space loss occurs due to signal spreading.

As a signal radiates outward from an antenna, it spreads into a spherical surface. The radiated power is thus spread across this surface, with the power level which is inversely proportional to the distance from the source antenna. Since antenna radiation must pass by and through solid objects, it suffers losses from reflection and absorption. Reflections from objects give rise to multipath distortion (fading). These signal can be constructive (adding to signal strength) or destructive (fading). The shape of the coverage area drives the selection of an antenna's azimuth pattern. Elevation pattern shapes are controlled to keep the maximum response at or slightly below the horizon for best far and near field coverage. The dimensions and height of a communication sector will determine an antenna's azimuth and elevation beamwidth requirements. These can be derived from the beam area formula for the approximate gain needed.

$G_{(dBi)} = 10 \log_{10}/29{,}000/....)$ antenna azimuth times the antenna elevation

Spacing in excess of 0.75 wavelengths from a large conducting surface leads to deep nulls. Antennas placed in the middle of a roof are prone to diffraction loss if they are not high enough to clear the roof edges.

SMART ANTENNA

Smart antennas are high gain, narrow beam antennas and use adaptive algorithms and DSP technology to increase the signal to noise ratio, and reduce the harmful effects of interference. Smart antennas utilize narrow beams of energy that are directional due to the intelligence in the logic section of the system. A system normally consists of an antenna array, associated hardware, and a controller. Using switched beam, and or adaptive array, the controller changes the response pattern.

Adaptive array

Adaptive array dynamically controls its pattern by using feedback to vary the phase and or amplitude of the signal received by each antenna element in order to optimize the received signal.

The use of smart antennas can yield more than 6 dB improvement in signal to interference performance as compared to sector antennas.

DUAL POLARIZATION ANTENNAS

The deteriorating effects of fading can, to so some degree is improved with diversity reception. The measure of this improvement is diversity gain. The usual method is to space two or more antenna's some distance apart, usually in the horizontal plane, and incorporate one of several techniques, such as maximal ratio combining, equal gain combining, or switch combining. This is called space diversity. Diversity gain relies on two signals that experience two different fading characteristics. Another way is to use two different polarizations in the same antenna. When an EM field is incident upon an object, the polarization of the signal is altered to some degree. A receiver can be rotated 360° in the vertical plane without any change in signal strength due to the depolarizing affects of multipath. This is the reason why dual polarized antennas can benefit a site. De-correlation in dual polarized antennas is directly related to the performance of the antenna. Optimum de-correlation occurs when the two polarization are orthogonal (perpendicular or crossed) and the antenna features high isolation between the two ports. Dual polarized antennas with slant +45° and slant -45° polarization give nearly identical radiation patterns.

TRANSMISSION ANTENNAS

Few transmission antennas are shown in Fig. 5.32 to 5.34. FM antenna range from simplest linearly polarized half-wave dipole right up through the circularly polarized Spearhead panel. A FM antenna panel is shown in Fig. 5.32. A rotatable logperiodic antenna is shown in Fig. 5.33. An antenna feeder system is shown in Fig. 5.34.

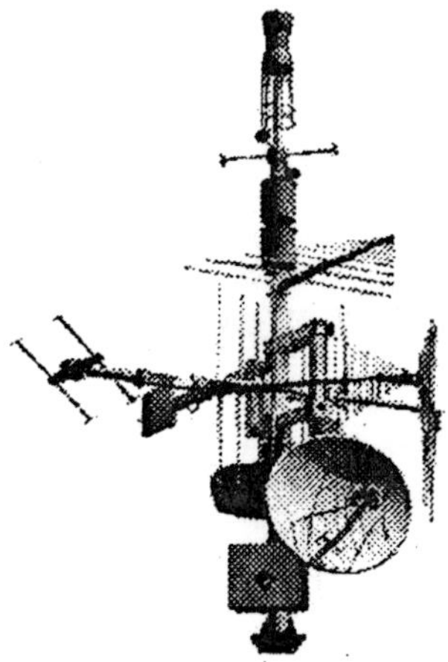

Fig. 5.32. FM Antenna Panel.

Fig. 5.33. Rotatable Logperiodic Antenna.

Fig. 5.34. An Antenna Feeder System.

Source: http://www.dxlc.com/nordx/fred.htmlThe SW site at Fredrikstad

CONCLUSION

In this chapter, the basic transmission system including the different type of modulations, transmitter components, brief description of DAB, HDTV, Satellite transmission system, propagation theory and transmission antennas have been described.

REFERENCES

1. Radioscientist On-Line , "The Sounds of a Spark Transmitter: Telegraphy and Telephony" at http://newton.otago.ac.nz:808/ ~ursi/spark.html

2. http://www.jct.ac.il/special/virtual/phone/phone11.html

3. FLIWH, General Information, "The Telefunken/Sayville WIRELESS" at http://asb.com/usr/w2g3jfz/radio/spark-transmitter.htm

4. http://robotics.eecs.berkeley.edu/~sastry/ee20/modulation/ node2.html.

5. http://www.bdcast.com/fm10b.html

6. HARRIS Communications at http://www.broadcast.harris.com/ transmission/radio/dx-series10.html#1

7. "Eureka 147/DAB system description DT/05959/III/C" December 1992.

8. "Additional and updated information on digital system for DAB, developed by Eureka 147 Consortium and supported by EBU", EBU Document WP 10 B/, WP 10-115/, December 1992.

9. D. Kramer, Proceedings of ITG sponsored conference on HDTV at SRG, 10 Nov. 1991.

10. ITU document 709.

11. Karen J. P. Howes, "Satellite in the Age of Interactivitiy", Via Satellite, Vol. 11, No. 2,pp. 114-125, Feb. 1996.

12. James Careless, "Ka-Band Satellites", Satellite, Vol. 11, No. 2 pp. 62-74, Feb. 1996.

6

RECEIVING SYSTEM

In this chapter the receiver portion of the broadcasting system shall be described. A receiver is a device that intercepts a signal, processes it, and converts it to a form useful to a person. The receiver will filter out unwanted noise signals, select the desired frequency, and amplify the weak signal. After amplification the signal is then demodulated. The process of demodulation is to strip the carrier from the modulated signal, thus recovering the original message. The chapter is not intended to give the circuit diagrams, since these are available in many standard books. Here, the thrust is on understanding the principles and also providing few tips on TV antenna system for home. The recent trends in the receiver technology, as a result of merger of broadcasting computer and telecommunication, shall also be described briefly.

PRINCIPLE OF RECEIPT SYSTEM

The modulated broadcast signal has to be finally demodulated to recover the information signal. The receiver does this. The receiver basically processes out the carrier by demodulation to reproduce the information signal Fig. 6.1.

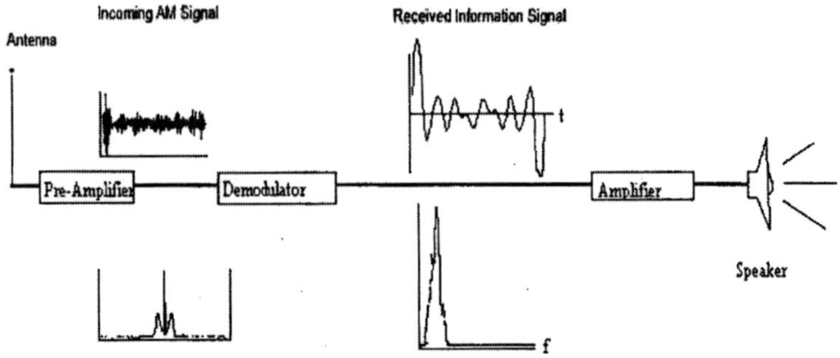

Fig. 6.1. A Schematic of Receipt System.

HETERODYNING

Heterodyning is a process where two frequencies are mixed together to produce a beat frequency, namely the difference between the two. Amplitude modulation is a heterodyne process since in this process, the information signal is mixed with the carrier to produce two side-bands. The side-bands occur at precisely the sum and difference frequencies of the carrier and the modulating signal. These are beat frequencies.

SUPERHETERODYNING

The term superheterodyne refers to creating a beat frequency that is lower than the original signal. Superheterodyning purposely mixes another frequency in the receiver, so as to reduce the signal frequency prior to processing.

SUPERHETERODYNE RECEIVER

Superheterodyning is simply reducing the incoming signal frequency by mixing. In a radio application, the AM or FM signal centered on the carrier frequency, is reduced to an intermediate frequency, called the IF (intermediate frequency). For practical purposes, the superheterodyne receiver always produces the same IF. To accomplish this the frequency being mixed into the signal is continuously varied so as to keep the difference the same. The block schematic of a superheterodyne receiver is depicted in Fig.6.2.

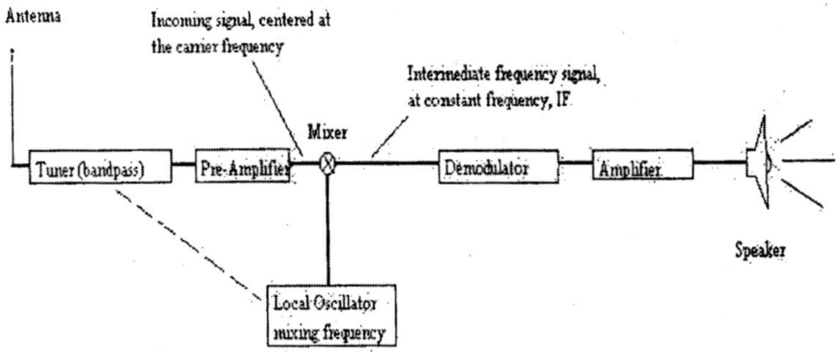

Fig. 6.2. Block Diagram of Superheterodyne Receiver System.

This is essentially the conventional receiver with the addition of a mixer and local oscillator. The local oscillator is linked to the tuner because they both must vary with the carrier frequency. For example, supposing a TV station at 235 MHz is to be tuned. The band-pass filter (which only permits signals in a small range about the centre frequency to pass) must be centered at 235 MHz (or slightly higher in SSB). The local oscillator must be set to a frequency that will heterodyne the 235 MHz to the desired IF of 452 KHz (typical). This means the local oscillator must be set to 234.448 MHz (or alternatively to 235.452 MHz) so that the difference frequency is exactly 452 KHz. The local oscillator must be capable of varying the frequency over the same range as the tuner; in fact, they vary the same amount. Therefore, the tuner and the local oscillator are linked so they operate together.

ADVANTAGES OF SUPERHETERODYNING

There are three main advantages of superheterodyning.

- It reduces the signal from very high frequency sources where ordinary components would not work.

- It allows many components to operate at a fixed frequency (IF section) and therefore they can be optimized or made more inexpensive.
- It can be used to improve signal isolation by arithmetic selectivity.

The ability to isolate signals, or reject unwanted ones, is a function of the receiver bandwidth. The band-pass filter in the tuner isolates the desired signal from the adjacent ones. In real life, there are frequently sources that can interfere with the desired signal. The frequency assignments generally prevent this. Frequently, the receiver bandwidth is some fraction of the carrier frequency. If the receiver has a bandwidth of 2 per cent and it is tuned to 850 KHz, then only signals within the range from 2 per cent above and below are passed. In this case, that would be from 833 to 867 KHz. Arithmetic selectivity takes that fraction and applies it to the reduced frequency (IF). For the fixed IF of 452 KHz, signals which are superheterodyned to the range of 443 to 461 KHz will pass. Taking this range back up into the carrier band, only carrier frequencies in the range of 841 to 859 KHz will pass. The local oscillator is set to reduce the 850 KHz to 452 KHz (i.e. must be set at 398 KHz). Thus, the 850 KHz is superheterodyned to 452 KHz. Any adjacent signals are also superheterodyned but remain the same above or below the original signal.

The superheterodyne receiver is more selective since it operates at a smaller frequency called arithmetic selectivity. Bandwidths that are expressed in percentage are smaller when the centre frequency is smaller. In cases where selectivity is important or the frequency is very high then superheterodyning can greatly improve performance.

SSB RECEIVER

In the SSB system, the receiver cannot output the signal as it is. It must first restore the signal to what it should be before demodulation. The receiver in a SSB system has its own carrier signal (from a local oscillator). The receiver schematic is given in Fig. 6.3.

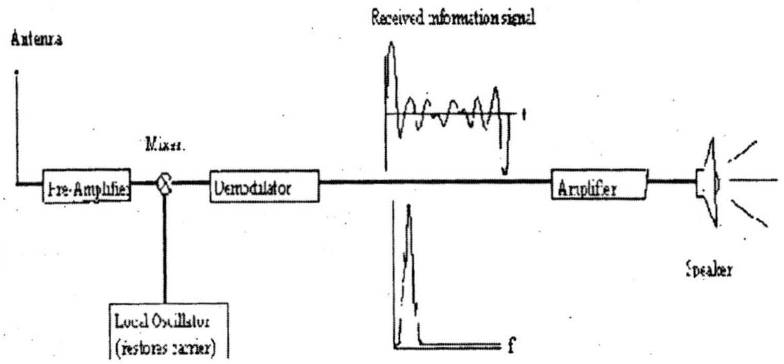

Fig. 6.3. Schematic of SSB Receiver.

By having its own carrier signal, the receiver makes the signal back into a shape that is sent by a conventional AM system. The remaining signal is processed normally.

BASICS OF FM RECEIVERS

The FM receiver has following stages.

RF INPUT STAGE

This stage takes output from antenna and amplifies this signal. This stage also reduces the image rejection problems.

MIXER STAGE

Mixer stage takes output from front end and by the use of heterodyning, outputs an IF frequency. The intermediate frequency required is balanced between the need for image rejection and the required bandwidth needed.

LIMITER STAGE

This stage removes any amplitude variations from the received signal.

DEMODULATION STAGE

The demodulation stage uses a discriminator circuit to demodulate or convert the received signal into audio voltages.

POWER AMPLIFICATION STAGE

This stage boosts the recovered audio to a level suitable to drive other devices.

ADVANCES IN THE RECEIPT SYSTEM

With the growing trend towards merger of broadcasting, communication and computers, the search for a receiver, which can double as a computer and TV, is ongoing. The development of digital interactive technology for domestic and personal use is arguably the phenomenon of the decade. Interactive television has been hyped and more recently disputed as the means by which digital services will be introduced to the home. Regardless of its title, Interactive television is set to address the core issues which continue to face new media: how will the cultural and business success of conventional television be linked to the possibilities of digital networks and information technology[1]. Future generation of television receivers would prove more than entertainment. It would pave the International Information Highway and build the common ground for communicating diverse high density, synchronous data. As digital video merges with the Internet, it can become the lingua franca of commerce, education, research and personal information. Few important R&D works are described here.

TVOT (TELEVISION OF TOMORROW) PROJECT

A TVOT (Television of Tomorrow) can be viewed as an extension of the Internet and the World Wide Web into entertainment or it could be viewed as an alternative model for television - Internet video. This TVOT would be capable of decoding of sound and picture, understanding the formats, archival retrieval, and a work station for personal creative expression. With this view, a forum by the name, "TVOT" has been set up as an internationally funded research consortium to address worldwide digital television and media. Short-term objective is to develop an open architecture television, and longer-term goal is a three-dimensional TV set.

NATIONAL INTERACTIVE COMMUNICATIONS PROJECT

The National Information Infrastructure (NII) program is to provide consumers throughout the United States, with an in-home system for entertainment, education, business, and even shopping, using the widely publicized digital interactive communications system[2]. The focus of the study is on the viewing requirements, consumer needs, technical considerations for inter-operability and scalability for a receiver set that can provide for HDTV viewing, computing and connecting to the information superhighway.

DEMONSTRATION AT NAB

For years, analysts have said PCs would not succeed as family entertainment devices until they exist in configuration that fit into the family entertainment centre and double as a TV. Now systems aiming to achieve that marriage are on the anvil. In NAB, products have been released which not only deliver PC staples such as application software, Internet access and CD-ROM titles, but which also function as stand-alone TVs. An MPC3 (Multimedia PC) compliant system can operate as an ordinary TV while the computer is of. The system can split the screen to display upto 12 TV channels simultaneously, or display a TV and PC image at the same time, by digitizing signals coming from cable or a VCR. It has video, computer and audio RCA input, and output jacks for integration into home theater system. There is also a 125 channel cable ready TV tuner and a broadcast channel lockout feature for parents to control children's viewing[3]. Several "ghost" cancelling systems have been proposed to enable modified television receivers to cancel multiple reflected signals and improve picture quality.

DIRECT TV LINK UP

Research is being carried out to provide video broadcasts and related interactive data services for personal computers using Digital Broadcast Satellite (DBS) System. Users will be able to download large files in seconds using the broadband delivery system. Microsoft plans to build support for TV reception directly into Windows 95. The direct TV set-top box will basically be shrunk down to a card that will fit in the PC and enable users to watch the regular video channels, either on the PC or on a connected TV, but data services will be available only on the PC. The service will also provide multimedia magazines with downloadable interactive programs and other services targeted at home consumers, including sports scores. There may also be tie-ins between video programming and related data services[4].

ANTENNA

ANTENNAS FOR TELEVISION

The quality of television reception depends on capturing a clear strong signal using a good quality antenna and feeding it into the television set. Outdoor antennas are preferred to the indoor variety whose performance is affected by wall insulation, plumbing, electrical wiring, roofing materials, etc. Antennas come in different sizes to cater to the gain requirements. The vertically polarized and horizontally polarized VHF/ UHF antenna systems for TV reception are shown in Figs.6.4. and 6.5. UHF Phased array system is shown in Fig. 6.6.

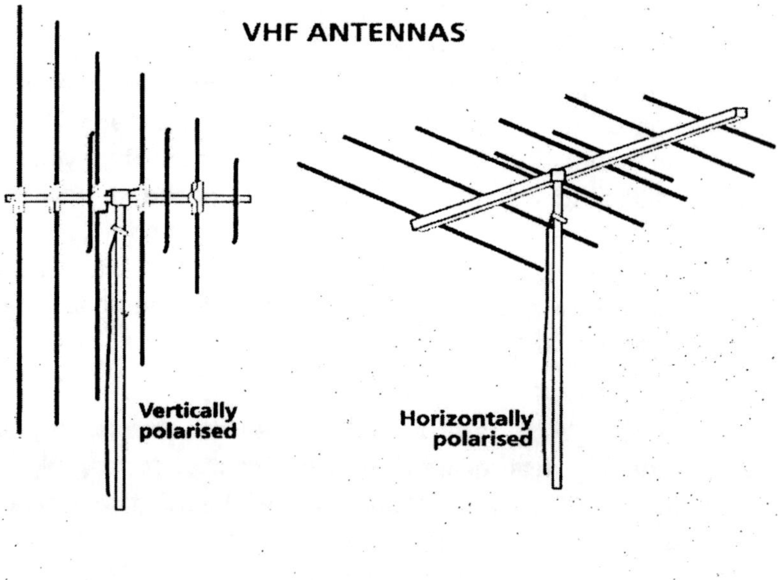

Fig. 6.4. Vertically / Horizontally Polarized VHF Antenna Systems for TV Reception.

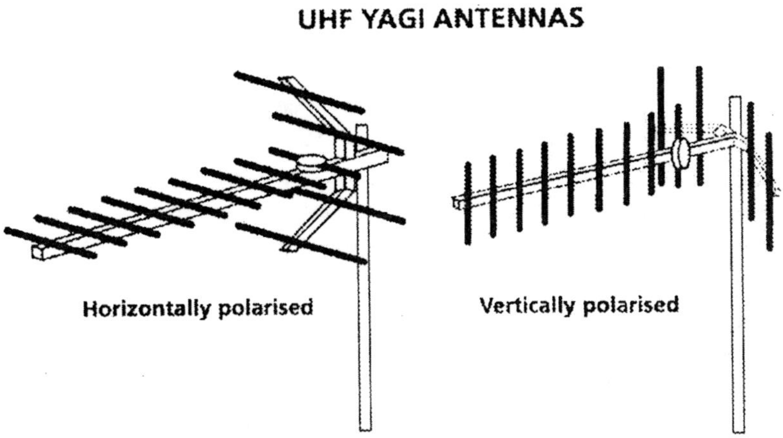

Fig. 6.5. Vertically / Horizontally Polarized UHF Antenna Systems for TV Reception.

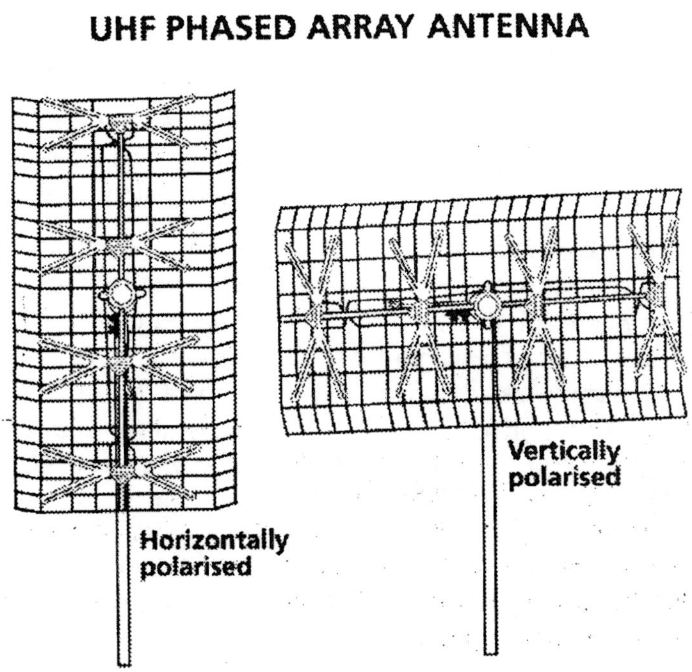

Fig. 6.6. UHF Phased Array System.

INSTALLING THE ANTENNA

The location of antenna is very important. It should be mounted on the side of the house with the clearest line of sight to the transmitting station. It should be ensured that there are no powerlines, large trees, hillocks or buildings obscuring the signal path. If the transmitting station is not visible, experiment by pointing the antenna to different directions may point the right direction to get the best picture on television screen. When mounting more than one antenna on the same pole, the higher band antenna should be placed on top (e.g. UHF antenna above a VHF antenna) and leaving a space of at least a meter between them[5].

If more than one television set is to be connected to the antenna(s), a splitter is used. A splitter takes the input signal from one cable and divides it between two or more outputs and then connects to the televisions. The use of splitter may sometimes cause the signal to deteriorate. This is because the signal strength reaching each TV gets diminished. This is overcome by using a masthead or distribution amplifier. The amplifier boosts the signal before it is fed into the splitter and ensures that signals of adequate strength are supplied to each connected TV set. Combined splitter/amplifiers are also available. Possible antenna connections are shown in Fig. 6.7.

FM radio broadcast signals sometimes interfere with television signals and cause annoying patterns over the picture. The problem occurs when high levels of FM signal are present at the input of a television receiver. Fitting a FM rejection filter to the TV's antenna input on the back of the set can often solve this. This filter is sometimes known as FM Trap. The FM filter enables televisions to reject FM signals thus preventing them from interfering with most television signals. However, they are unable to protect Band 2 channels (i.e. TV channels 3, 4 and 5) as these channels share part of the Radio Frequency Spectrum with FM radio. Using a FM filter would filter out the signal from TV channels 3, 4 and 5 along with the unwanted FM radio signal.

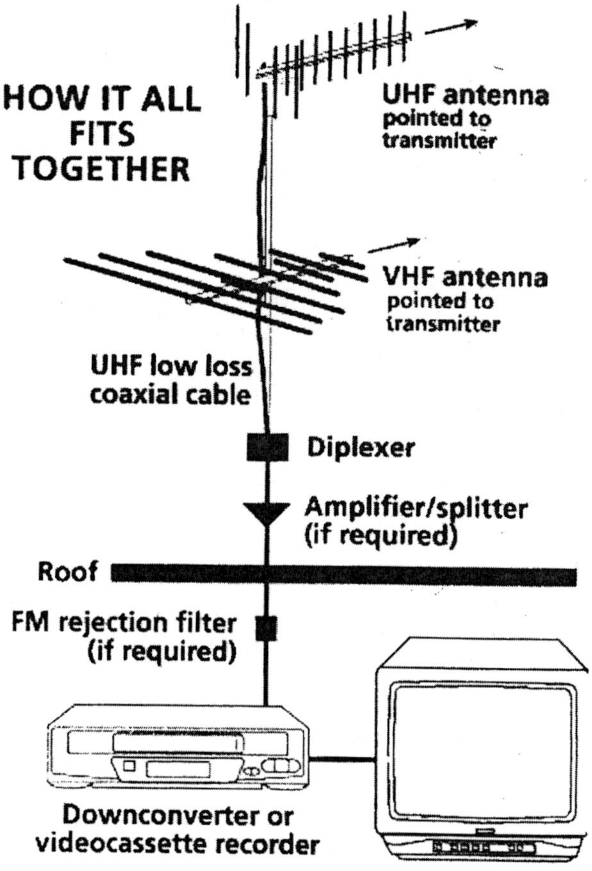

Fig. 6.7. Possible Antenna Connections.

CONCLUSION

This chapter has described briefly the principle of broadcast receipt systems, new advancement in the receipt systems in view of the merger of broadcast, computer and telecommunication systems, and also the TV antenna connection for VHF/UHF reception.

REFERENCES

1. "Montreux Symposia", *World Broadcast News,* Vol. 19, No. 5, pp. 52-56, May 1996.
2. Craig, J. Birkmaier, "A Commentary on Requirements for the Interoperation of Advanced Television with the National Information Infrastructure " at http://www.eeel.nist.gov/advnii/birkmaier.html.
3. Britton Manasco, "Open Architecture DBS and the era of Customized Television", Via Satellite, Vol. 11, No. 6, pp. 42-48, June 1996.
4. James Careless, "Surfing among the Starrs", Via Satellite, Vol. 10, No.12, pp. 30-34, Dec. 1995.
5. "Improving Your Television Reception" at http://www.cybercomm.nl/~hpl/anten.htm

7

MEASUREMENTS, CHARTS AND TEST SIGNALS

This chapter will cover the equipment needed to measure the video signal. To a certain extent eyes can be trusted to provide a good general evaluation of the quality of a TV picture. However, in most applications there are multitudes of variables that can affect visual judgment. Items such as ambient lighting near the television monitor or accuracy of monitor set up vary between environments. As such, two specially built oscilloscopes are used to provide consistency in measuring video signals. To measure luminance, the waveform monitor is typically used. To measure the colour or the chroma element, the vectorscope is typically used. While the waveform monitor offers some measurement of the colour, the vectorscope is usually the instrument of choice for colour evaluation.

A large variety of test and monitoring techniques have evolved after the introduction of commercial broadcasting of monochrome television, and colour television service. This includes use of test signals, charts, and procedures. This involves the generation of pictorial information in the form of representative electrical signals, transmission of this information through a specified channel, and eventual display of the information in order to make the perceived image appear the same as the perceived original scene. These methods have been developed to fulfil the technical as well as regulatory need to monitor and measure the performance of all segments of the total television system, both the individual components and the total signal path. A brief summary of some of the most notable and widely utilised test signals and charts are described in this chapter.

A variety of test charts of appropriate geometric formats have been developed to measure the quality of the transfer from optical information to electrical signals by the camera at one end and the picture monitor, or display, at the output end of the system. In addition, a wide variety of electrically generated test signal formats are utilized that may be injected into the system at appropriate points for both objective and subjective performance checks.

THE WAVEFORM MONITOR

Waveform monitor and Vectorscope are used to see the various elements of the composite video signal. A waveform monitor is an oscilloscope that is designed to display the luminance, chroma and sync levels of a video signal. Since brighter video levels are made up of higher voltages, this display device is useful for setting, adjusting and monitoring video levels.

The waveform monitor displays the black and white (luminance) video signal information. A scale, called the graticule, is used to make measurements on the waveform monitor. The scale is based on IRE (Institute of Radio Engineers) units. An IRE unit is 1 per cent of the difference between the peak white (maximum) level and the blanking (absolute black) level. The IRE scale can be found on the vertical axis of the waveform monitor. Signals that are outside the range of 100 and 7.5 IRE lose all detail so that if a light coloured face is at or above 100 units, it will be washed out. A dark coloured face below 7.5 units will be so dark as to have no detail. 100 is called the white peak and is called Reference White, because it is where the brightest white in the picture should peak produced by the brightest allowable item in the scene. Video peaks will vary according to what is the brightest in the scene. If the production planning process considers the ideal situation for camera set-up, they will have included something that is TV white in every shot. The line marked 7.5 is called the black peak because it is where the black in the picture should peak. The term black level defines the actual position, measured in IRE units, on the waveform of what is black in the scene. Faces in the +50 to +80 range are generally considered to be properly exposed. Fig. 7.1 shows the colour bars on a waveform monitor.

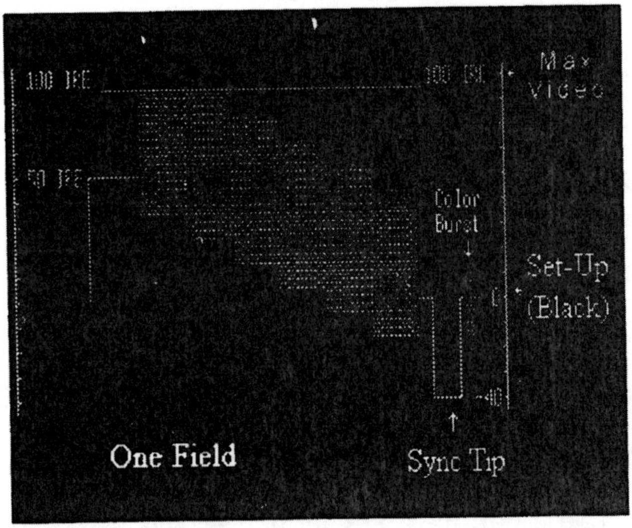

Fig. 7.1. Colour bars on a waveform monitor.

WAVEFORM MONITOR MEASURES VOLTAGE LEVELS

The actual voltage is measured on the waveform monitor. From the 100 unit level to the -40 unit level is exactly 1 volt. Between 0 and 100 units, the measured voltage is exactly 0.714 volts. Between 0 and -40 is a value of 0.286 volts. Adding the two together makes I volt. This is the level at which video signals are distributed between cameras, VTRs, switchers, television monitors, field cameras and portable VCRs. On the waveform display there is a space that separates each line. This is called horizontal blanking, where the scanning beam is turned off during horizontal retrace. Areas defined as the front porch and back porch identify the blanking width. The colour information for the video signal is contained in the colour burst signal located next to the back porch (Fig. 7.2). The colour burst normally sits between -20 IRE to +20 IRE units.

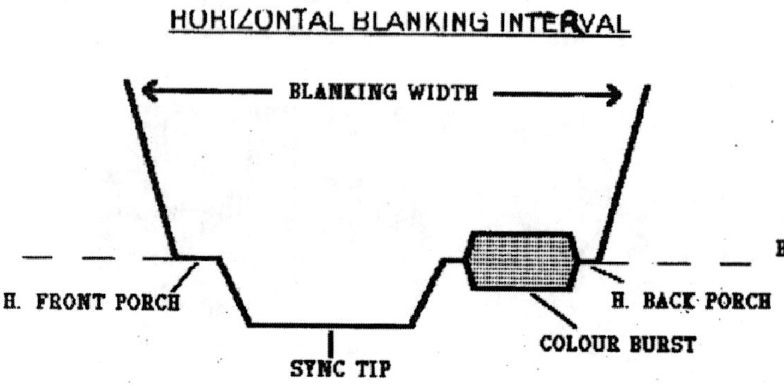

Fig. 7.2. Colour Burst.

ZEBRA PATTERN

When the video is shot in the field the camera operator uses a level indicator built into the camera to assist in finding the proper video level. The most common of these level indicators is the Zebra pattern. It alerts the camera operator to correct light level by spreading a striped "Zebra" pattern over correctly exposed portions of the video image. This Zebra pattern is only displayed in the camera viewfinder and is never recorded on videotape. The Zebra pattern requires a certain amount of personal judgement in deciding correct video levels.

Colour Bars

Colour bars are the standard video reference tool used in television. Colour bars are used as a reference and contain correct luminance and chrominance (colour levels). Colour bars are also recorded at the beginning of each videotape. This allows videotape technicians and editors to establish correct videotape playback levels using the colour bars as a reference when the tape is being edited or copied. Colour bars built into cameras are used when recording in the field. Camera colour bars are also used when synchronizing multiple cameras in a studio situation. Modern colour bars contain the primary and secondary colours, produced by the colour additive system used in television. They also contain a reference white signal, a reference black signal, the monitor set-up bars and I and Q reference signals. There are three basic types of colour bars in use today.

EIA CHARTS

Recognition that cameras and related equipment required standardized calibration procedures led the EIA (Electronic Industries Association) to design an initial series of five television test charts. These charts were developed by the EIA TR4.4 Studio Facilities Committee and have been in wide use over the years. The five test charts are listed below:

1. *EIA Color Registration Chart.* This chart is used to aid the alignment and test the registration of the three-colour pick-up live and film colour television cameras. The very fine horizontal and vertical black lines on a white background permit accurate alignment of the optical and electrical camera systems.

2. *EIA Logarithmic Reflectance Chart.* This chart has two crossed, nine-step gray scales on a 12 per cent gray background, with the scale steps showing a logarithmic reflectance progression from 3 to 60 per cent. Between the scales is a black level switch for absolute black calibration. The chart is used for checking and adjusting of the camera transfer characteristic.

3. *EIA Linear Reflectance Chart.* This chart is the same as the Logarithmic Reflectance Chart except that the nine gray scale steps have a linear relationship to each other.

4. *EIA Resolution Chart.* This chart is used to measure the resolving power of the system. The chart is also useful in checking scanning size, linearity, aspect ratio, focus, focus uniformity, shading, low-frequency phase shift, resolution, and transient response. The chart is televised by the camera and displayed on a monitor. When adjusted according to instructions relative to the scanned area (television raster dimensions), the horizontal and vertical resolution wedges cover the range from 200 to 800 television lines. Gray scale overlays are available for the indicated

sections of the chart. These gray scales are designed to provide logarithmic reflectance relationships. Both the horizontal wedges (vertical resolution) and the vertical wedges (horizontal resolution) are calibrated in both frequency and television line-number increments.

5. *EIA Linearity (Ball) Chart.* This chart is used to test geometric distortion of the camera chain. Comparing on a monitor, two superimposed patterns, one generated electrically and the other by viewing the television chart with the camera does this. The electrical frequencies required to match the chart pattern are 315 KHz for horizontal and 900 Hz for the vertical directions. The inner and outer diameters of the circles represent 1 and 2 per cent of the picture height, respectively.

Three additional EIA Test Charts, given below, were made available in 1982.

1. *Multiburst Chart.* This chart consists of vertical alternate black and white bars whose repetition rates produce signal frequencies of 0.5, 1.0, 2.0, 2.5, 3.0, 3.5, and 4.0 MHz when the scanned area is adjusted according to instructions. It provides a measure of the operating modulation transfer function (MTF) (frequency response) of the camera chain and subsequent signal-processing apparatus.

2. *White Pulse Chart (Window Chart).* The chart is used to adjust image enhancement apparatus and check the number of times the system rises and transient response. The pattern consists of a white box against a black background to produce a full dynamic range square pulse on a waveform monitor.

3. *Colour Calibration Chart.* This chart consists of three-horizontal bars on a black background. The centre bar is constant at 60 per cent white level. The other two bars show six saturated colours representing the hues of the NTSC primaries and their complements. On a vectorscope, these colours fall within the designated tolerance boxes when the system is properly

calibrated. This process may match the colour performance of two or more cameras focused on the same chart.

A great variety of electronically generated equivalents of these charts as well as many other pattern formats for special test purposes are available from numerous manufacturers. These pattern generators produce idealised composite signals to be injected into the television system at the appropriate point, independent of the actual optical/ electrical camera equipment. A number of test signal waveforms have been initiated over the years for purposes of testing the performance of colour television processing apparatus. Few colour bars are described below.

EIA COLOUR BARS

The first type, standardized in 1967, conform to the EIA (Electronic Industries Association) standards. They are known as full-field bars, and also RS-189 bars. These are shown in Fig. 7.3. In 1976, these bars were changed by the EIA, and became the RS-189A standard. More commonly known as split-field bars, they added a reference white signal, a reference black signal, and I and Q reference signals. They were re-designed for easier set-up on a vectorscope. These are shown in Fig. 7.4.

SMPTE COLOUR BARS

The EIA split field bars were weak, especially in the area of monitor set-up. In 1978, Society of Motion Picture and Television Engineers (SMPTE) adopted the third type of colour bars that addresses monitor set-up. This was accomplished by adding a tric of black bars to the black reference signal. The idea is to adjust the brightness on the monitor until the bar on the left cannot be seen. If it can be seen, brightness is considered too high. If the right hand bar cannot be seen then the brightness is too dark. SMPTE bars have become the industry standard. The SMPTE colour bars are shown in Fig. 7.5.

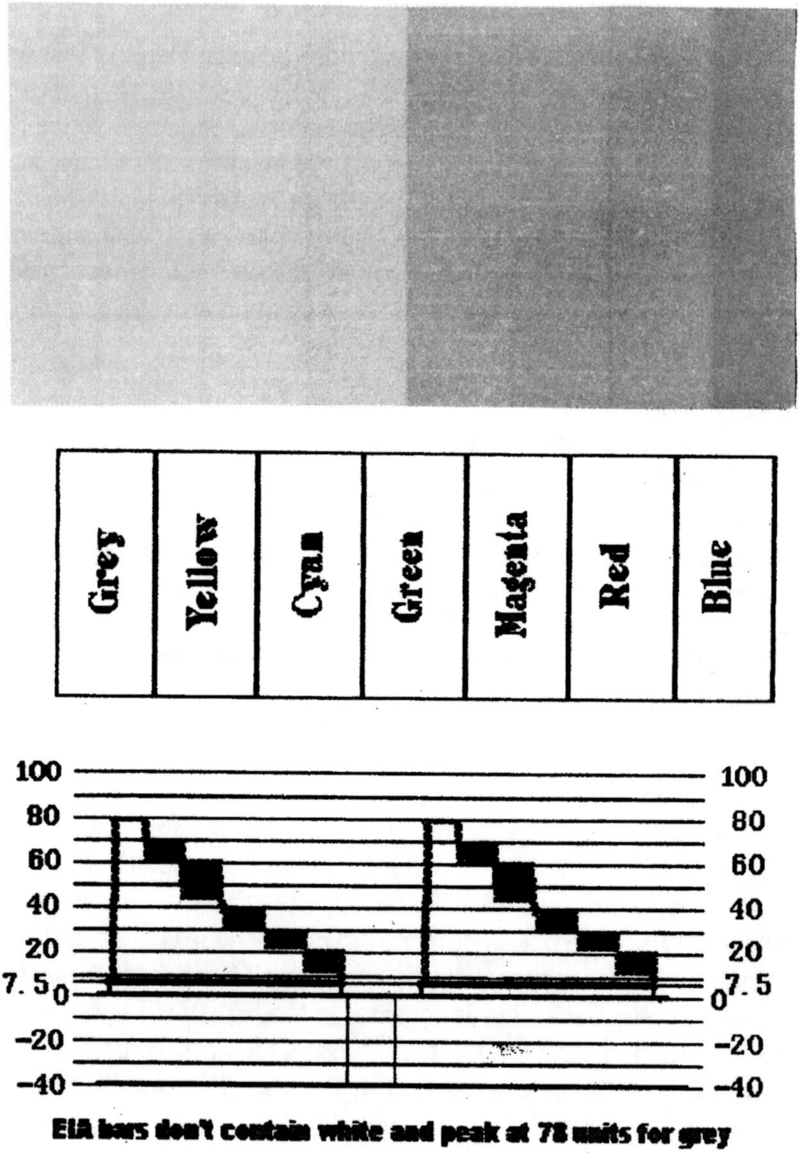

Fig. 7.3. EIA Full-field Bars (RS-189).

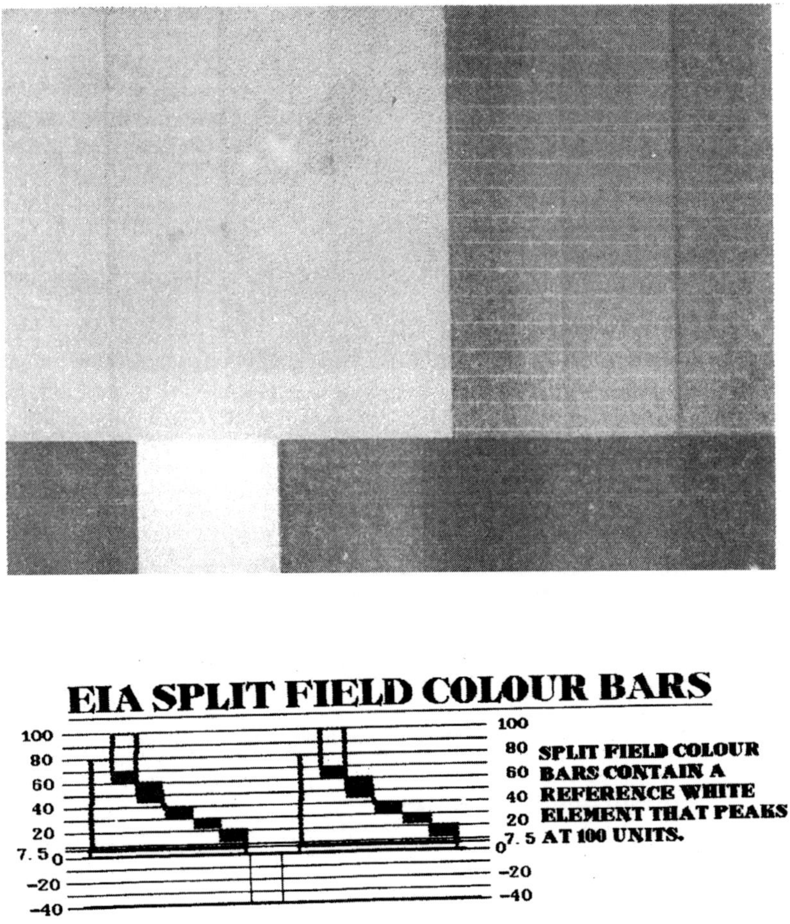

Fig. 7.4 EIA Split colour bars (RS 189-A)

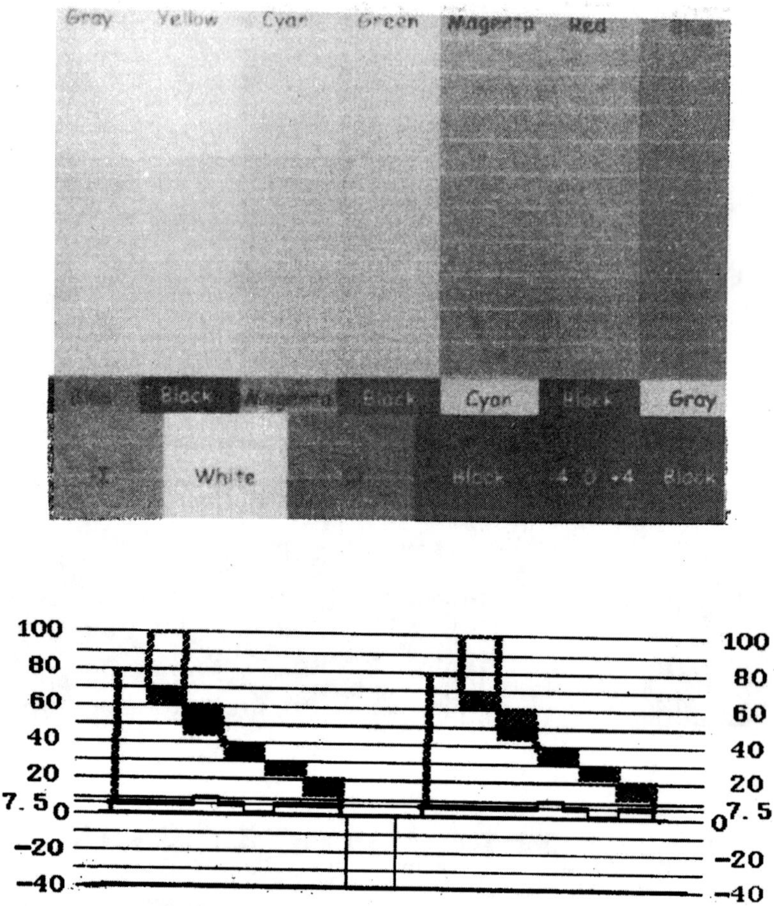

Fig. 7.5 SMPTE Colour Bars

TEST CARDS

Few test cards used by different nations at various times are given in Appendix II.

THE VECTORSCOPE

Typically used in conjunction with the waveform monitor, the vectorscope displays and measures the chrominance (colour) of the video signal. The scale of the vectorscope is a circle overlaid with the colour amplitude and phase relationship of the three primary colours (red, green and blue). In the centre of this circle graph is the luminance (black and white) value of the signal. Through this centre point, three axis represent the primary colours (Fig. 7.6).

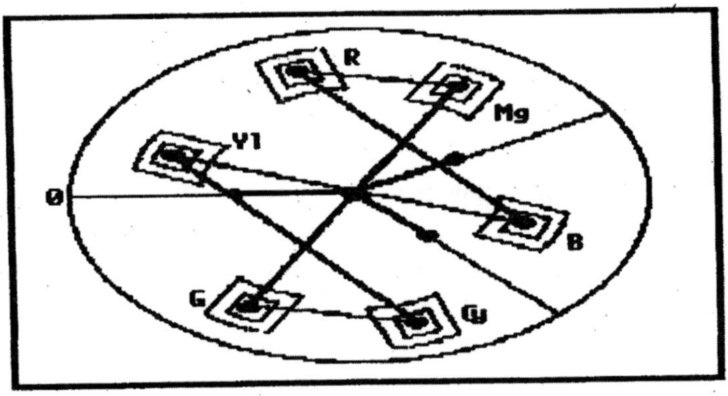

Fig. 7.6. Vectorscope Signal.

If a camera is pointed at a white card, the vectorscope shows a flat line that extends along the base reference of the vectorscope. If this line is closer to any of the three axis, the white card is not recorded as pure white, but with a tint of colour. To record the white as white, the cameraman must use the camera's white balance control. The camera is also adjusted internally with the red and blue gain control. On the vectorscope this is adjusted until the signal on the scope are dead centre and not favouring red, blue or green.. In the on-line edit session, the vectorscope determines the proper colours through the use of colour bars. The standard procedure is for the cameraman to record 60 seconds of colour bars at the beginning of every tape. This insures that when edited, the colours are the same from tape to tape and from any effects that are generated in the on-line editing session.

THE PROCESSING AMPLIFIER

This device is called a proc amp, can modify both the chroma and luminance values of the video signal. The proc amp is an invaluable tool for correcting the colour and luminance elements of a video signal. The proc amp is used with both a waveform monitor and vectorscope to ensure the changes. By adjusting colour, contrast and the other controls, the appearance of the picture on display device can be changed without changing the source of that picture. However, the proc amp allows the change of the actual signal rather than just the display. For example in an edit session, proc amp's chroma gain control can be reduced for getting less colour saturation on master edit tape.

CONCLUSION

This chapter briefly describes the test signals, charts and procedures for measurement. Waveform Monitors and Vectorscopes are well established instruments used in measuring video. Most scope manufacturers now offer combination units that act as both waveform monitors and vectorscopes. On the leading edge are waveform monitors and vectorscopes that are computerized and do not employ a cathode-ray tube. Currently there are scopes available that use liquid crystal displays and key the waveform or vector display over the video image. The computerized versions promise improved reliability and easier set-up.

8

BROADCASTING IN INDIA

Broadcasting in India during last seventy-seven years has seen a tremendous growth in terms of its reach and impact. Starting from one medium wave station set-up by the Radio Club of Bombay in June1923, India has at present a massive network of 197 radio stations with 145 Medium Wave, 55 Short Wave and 105 FM transmitters for sound broadcasting #. On the TV side, it has 47 programme generating centres, 93 High Power and 958 Low Power transmitters*. This chapter traces the history of this growth and the future of broadcasting in India.

As on 1.7.99. * As on 1.7.99.

INTRODUCTION

The history of radio broadcasting in India dates back to June 1923 when the Radio Club of Bombay started its first station. In 1927, Government of India entered into an agreement with a company named Indian Broadcasting Company to start broadcasting from radio stations at Bombay and Calcutta. In 1935, the Indian State Broadcasting Service was renamed as All India Radio (AIR). At the time of independence, All India Radio had 6 stations at Delhi, Calcutta, Bombay, Madras, Lucknow and Tiruchirapally. At that time, AIR's 6 medium wave transmitters and 12 short wave transmitters covered 2.5% of the area and 11% of the population of India. Soon after independence a scheme for development of broadcasting, known as the Pilot Project, was drawn up. Subsequently in 1951, with the commitment of the country to centralized planning, AIR's expansion and development was included in the Government's successive Five-Year Plans.

The various options available to domestic broadcasting in terrestrial mode of transmission are MW, FM and SW. AIR used SW along with MW for economic consideration although SW provides only a second grade service. However in a vast country like India, it served the purpose reasonably well since during 60s and 70s, SW spectrum was much less crowded and man-made noise was also low. In India, broadcasting remained confined to AM (except for setting up of FM transmitters at four metropolitan cities) up to to 1970s. By 1980, medium wave provided coverage to 86 per cent of the area and 95 per cent of population, with SW mainly being a back-up service.

SERVICES PROVIDED BY AIR

The power of broadcast service has been harnessed for socio-economic development of the country through 'Home Service' broadcasts. For home services, All India Radio follows a three-tier system of broadcasting to cover the nation. This consists of national, regional and local level broadcasting. India being a multi-lingual, multi-cultural country, Home services broadcast 2300 hr. of programme in 24 languages and 146 dialects everyday. The cultural renaissance and the country's point of view is put before the world through the 'External Services' broadcasts. The external services division was galvanised into action with the explosion of World War II. Among the most noteworthy contributions to the war effort was a daily five minute programme called 'Counter attack' (*Jawabi Hamala* in Hindi). It successfully countered the forceful Nazi propaganda through imagination and powerful presentation. As of date, this service caters to the international audiences covering 54 countries in five continents. External services provide a daily broadcast of 73 hours in 16 foreign and 8 Indian languages through 19 transmitters at 9 transmitting centres. The broadcasts are sprinkled with the news services through 288 National and Regional new bulletins in 64 languages/dialects.

The set-up of AIR consists of 197 broadcasting centres, 55 short wave transmitters, 145 medium wave Transmitters, 105 FM

Transmitters. The growth of stations is shown in Figure 8.1. A few important milestones are given in Table 8.1.

Table 8.1: Important Milestones

1952 : National Programme of Music
1953 : National Programme of Talks (English)
1954 : Radio Sangeet Sammelan
1956 : National Programme of Plays
1957 : Vividh Bharati Service
1967 : Commercials on Vividh Bharati
1968 : National Programme of Talks (Hindi)
1977 : FM Transmission from Madras
1984 : Local Radio Station at Nagercoil
1985 : Hourly News Bulletins
1988 : FM Stereo Transmission
1989 : FM Broadcasting Centre at Kottagudam
1996 : Test Digital Audo Broadcasting
1999 : Digital T.V. Transmission.

BROADCASTING CENTRES

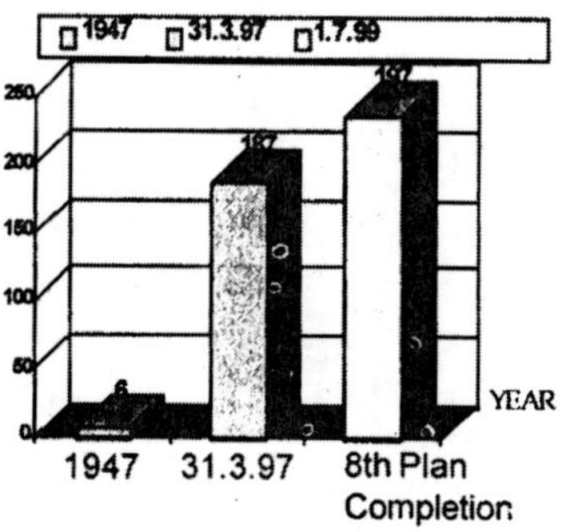

Fig. 8.1. AIR Network.

HOME SERVICE BROADCAST

AM Broadcasting

Broadcasting in medium wave band provides first grade quality service to listeners covering large areas comprising primary service zone around the transmitters during day-time. As such during 60s and 70s, the main thrust was on MW radio stations. Medium wave propagation has in addition to ground wave component, a skywave component which is capable of providing coverage to long distances during the night-time. However, the night-time capability of medium wave also causes interference to the other stations operating on the same frequency or adjacent frequency. With the increase in number of channels and man-made /industrial noise, it was observed in 1980s that the population coverage had gone down from 95 per cent to 60 per centduring night-time.

The international fraternity of broadcasters became alive to the interference problem of medium wave in the early 70s itself and therefore ITU decided to prepare a LF/MF plan. Under this plan, AIR obtained 149 medium power/ high power day and night assignment on 114 frequency bands (channels) at 107 places. Besides these, 352 one kW assignments were obtained. Out of the high power assignments allotted to AIR, the power of 12 medium wave transmitters was upgraded during early 80s and 29 medium wave transmitters during late 80s and 90s. The growth and coverage of medium wave transmitters are given in Figures 8.2 and 8.3.

GROWTH OF AIR NETWORK MEDIUM WAVE TRANSMITTERS

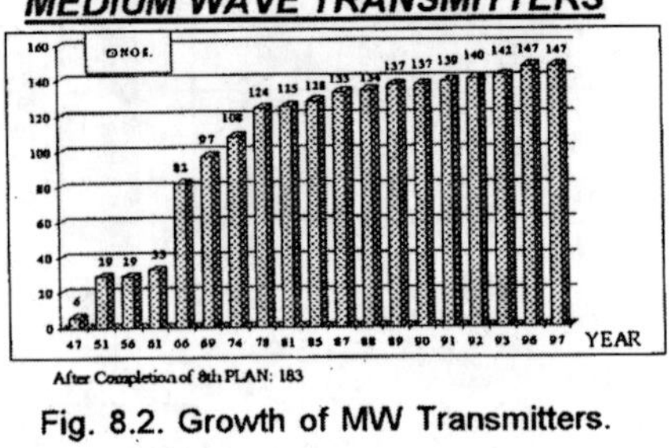

After Completion of 8th PLAN: 183

Fig. 8.2. Growth of MW Transmitters.

GROWTH OF AIR NETWORK
COVERAGE OF MEDIUM WAVE

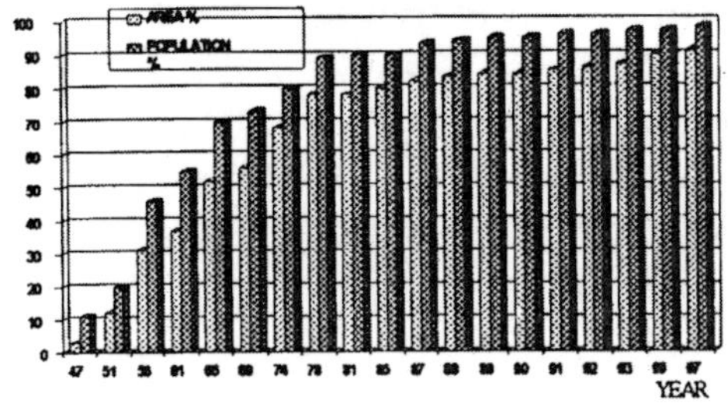

Fig. 8.3. Growth of Coverage.

Short wave broadcasting, in India, was initially used to provide service to uncovered areas like hills and also as a radio link for centrally relayed programmes like news, commentary, national broadcasts etc. Even with all the associated problems of ionospheric propagation viz. fading and deterioration of signal due to solar emission, sun spots, solar flux/flares, etc., its wide reach made it popular during sixties. Short wave service, even today works as a back-up service to Satellite links for home services transmitters. As on today, the number of SW transmitters are shown in Figure 8.4.

GROWTH OF AIR NETWORK
SHORT WAVE TRANSMITTERS

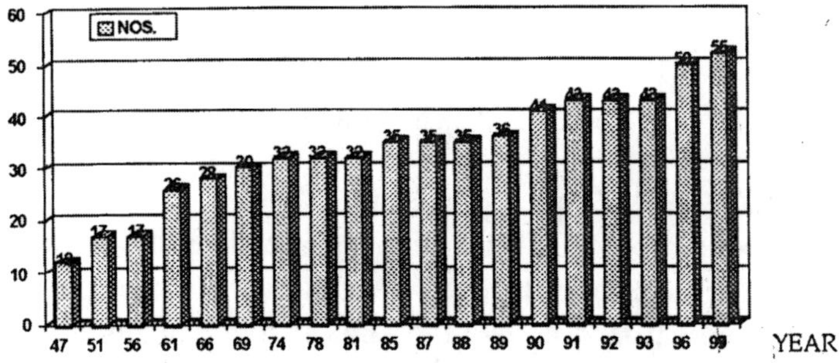

Fig.8.4. Shor twave Network Growth.

FM broadcasting

FM broadcasting is characterized by uniformity of reception both during day and night. It also provides noise free listening besides the advantages of repeatability of frequency, high fidelity, capture effect and possibility of introducing stereophonic transmission and RDS (Radio Data Service). Due to these advantages, AIR introduced FM transmission in a big scale during 1990s. The growth of FM transmitters is given in Figure 8.5.

GROWTH OF AIR NETWORK F.M. TRANSMITTERS

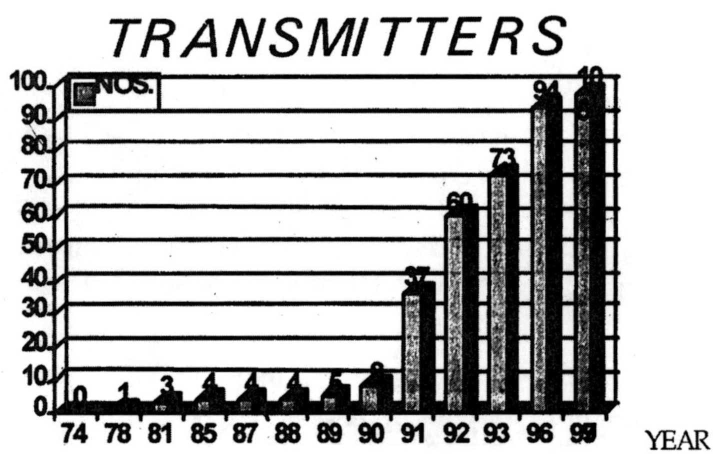

Fig. 8.5. Growth of FM Network.

EXTERNAL SERVICES

For external services, the technological options till now had been confined to SW broadcast or MW transmissions for those countries that may be situated across the sea. The transmitters used for external services of AIR are given in Table 8.2.

Table 8.2 Transmitters used for External Services

Short Wave Transmitters		
ALIGARH	4 Number	250 kW
BANGALORE	6 Number	500 kW
DELHI	2 Number.	250 kW
GORAKHPUR	1 Number	50 kW
PANAJI	2 Number	250 kW
Medium Wave Transmitters		
CALCUTTA	1 Number	1000 kW
JALLANDHAR	1 Number	100 kW
RAJKOT	1 Number	1000 kW
TUTICORIN	1 Number	200 kW

TELEVISION BROADCASTING

All India Radio made a modest beginning in this field on September 15, 1959. One hour long twice a week experimental television service was provided at New Delhi which was available within a radius of 25 Km from the Broadcasting House. The regular one hour a day service commenced on August 15, 1965. The second television station in the country was set up at Mumbai, 13 years after the inception of the experimental service at New Delhi. By 1982, only 25 per cent of the then Indian population could receive television programmes. Television broadcasting was entrusted to a new department on April 1, 1976 by an Act of Parliament. This department is known as TV India (Doordarshan)

SATELLITE INSTRUCTIONAL TELEVISION EXPERIMENT (SITE)

The Department of Space and Doordarshan made use of communications satellite ATS-6 lent by NASA in 1975-1976 to study the impact of television on the Indian people. Television posts capable of receiving programmes directly from this satellite were set-up at

2400 locations in six states. SITE was an important landmark in the history of Indian television as it laid the foundation of satellite television service in the country.

GROWTH OF TELEVISION IN INDIA

Doordarshan (the Indian National Television Network), is now a part of Prasar Bharati—the autonomous Broadcasting Corporation of India. Doordarshan, one of the largest broadcasting organizations in the world had a modest beginning in 1959. By end of June 1999 Doordarshan operate 19 channels, has a network of 1051 transmitters, puts out over 1,400 hours of programmes every week. The terrestrial signals of Doordarshan can reach 87 per cent of the country's 950 million people. Doordarshan services at a glance is given in Table 8.3. The details of Doordarshan transmitters are given in Table 8.4. Growth of Doordarshan transmitters has been shown in Fig. 8.6. The progressive growth of Doordarshan coverage is shown in Fig. 8.7.

Table 8.3. Doordarshan services at a glance

(As on 1.07.99)

Channels	19
Studios Centres	47
Transmitters	1051
Satellite Transponders in use	19
Area Covered (per cent)	72.9 %
Population Covered (per cent)	87.6 %

Table 8.4. .Doordarshan Transmitters

	DD-1	DD-2	Others	Total
High Power	84	9	-	93
Low Power	660	42	3	705
Very Low Power	231	4	-	235
Transposer	18	-	-	18
Total	993	55	3	1051

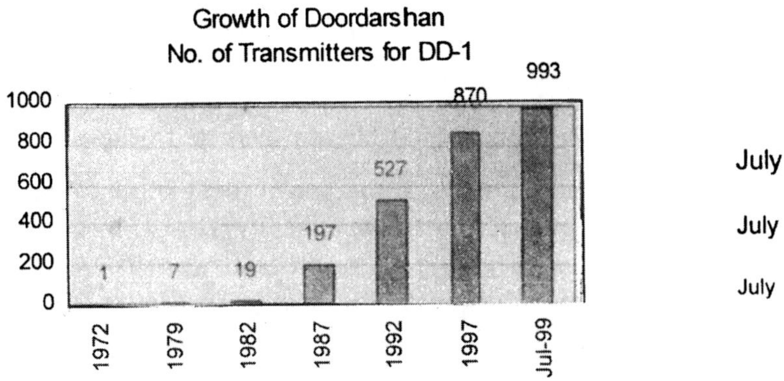

Fig.8.6. Growth of Doordarshan Transmitters.

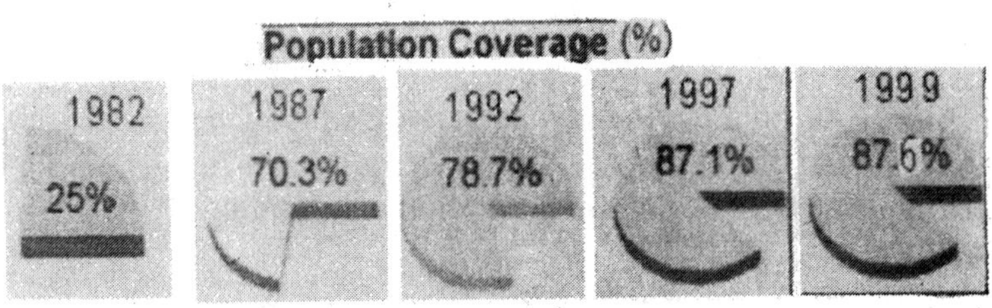

Fig.8.7. Growth of DD Coverage.

DD Channels

Doordarshan operates 18 channels—two All India channels, 11 Regional Languages Satellite Channels (RLSC), four State Networks (SN) and an International Channel. On DD-1 National programmes, Regional programmes and local programmes are carried on time-sharing basis. Metro Entertainment channel carries Network entertainment programmes from Delhi and also single-Metro programmes from the four metro cities. The RLSC have two components—the Regional service for the particular state relayed by all terrestrial transmitters in the state and additional programmes in the regional language in prime-time and non-prime time available only through cable operators. The four Hindi language states have state-networks which enable the relay of the programmes originated from the state capital by all the transmitters in the state. In Jammu and Kashmir there is provision to relay the news bulletins of DDK Srinagar by all transmitters in the state. DD-India is targeted to viewers who may be living outside India but are interested in India.

In addition, there are a number of local kendras relaying programmes on a single transmitter. In Delhi two low power Transmitters operate to bring live the proceedings of the two Houses of the Parliament. In Srinagar there is another Low Power Transmitter relaying programmes of interest to the people in the Kashmir valley. The details of Doordarshan Channels are given in Table 8.5.

Table 8.5. Doordarshan Channels

DD-1	All India	Primary service
DD-2	All India	Metro Entertainment channel
DD-4	RLSC	Malayalam
DD-5	RLSC	Tamil
DD-6	RLSC	Oriya
DD-7	RLSC	Bengali
DD-8	RLSC	Telugu
DD-9	RLSC	Kannada
DD-10	RLSC	Marathi
DD-11	RLSC	Gujarati

DD-12	RLSC	Kashmiri (DDK Srinagar)
DD-13	RLSC	Assamese and Languages of North East
DD-14	SN	Rajasthan
DD-15	SN	Madhya Pradesh
DD-16	SN	Uttar Pradesh
DD-17	SN	Bihar
DD-18	RLSC	Punjabi
DD-	India	International service

TERRESTRIAL AND SATELLITE TELEVISION CHANNELS

Doordarshan is utilizing the VHF and UHF band of frequencies for terrestrial transmissions. The satellite television service is in 'S', 'C' and 'extended C' band of frequencies. Doordarshan utilizes the transponders in the INSAT system of communication satellites for supporting its 19 services. Ku-band is yet to be harnessed for the transmission of programmes directly into the homes (DTH). Programmes are transmitted in the analogue form and are free to air.

DD-INDIA (INTERNATIONAL) CHANNEL

The International channel of Doordarshan started on 14th March 1995 on a transponder hired from Zee TV. Initially the telecasts were for a duration of three hours a day for five days in a week. When Doordarshan acquired a transponder on PAS-4, a daily service was introduced and transmission time was extended to four hours a day in July 1996. In November 1996, transmission time was further extended to 18 hours a day comprising nine hours capsule and its repeat telecast from 6.00 a.m. to midnight. The signal of DD-International can be received in countries of South Asia, the Gulf, Middle East and Europe via the PAS-1 satellite and in North America via the PAS-1 satellite.

The programme mix aims at providing an update on Indian social, cultural, political and economic scene with trade and export being major areas of interest. The channel focuses on news, current affairs, stock market update and discussions on issues of international significance.

CABLE TELEVISION

Cable television transmission network started mushrooming in India in a bigway in 1993. The cable television networks were regularized through a comprehensive act of parliament in 1995.

PROGRAMMES

Doordarshan has three-tier programme service—the national, regional and local. Its programmes fall under three distinct categories of education, information and entertainment.

Currently Doordarshan originates 1300 hours of programmes per week on its various terrestrial and satellite television channels through 42 programme production centres. To meet the demands of globalization, it has launched an International channel. Doordarshan programmes are currently being watched in 50 countries in Asia and parts of Africa and Europe. This source will soon be made available to most parts of the globe using international satellite systems like Intersat and PanAm Sat, etc.

Doordarshan started the first education television service at New Delhi in 1961. ETV Programmes are currently produced and telecast from Chennai, Delhi and Mumbai as well as from several other centres. University Grants Commission's countrywide classroom programme is also being supported on the national channel of Doordarshan with a view to putting quality education within the reach of small villages and towns. Syllabus based programmes for the students of Indira Gandhi National Open University are being telecast by Doordarshan.

Doordarshan telecasts news bulletin in English, Hindi as well as regional languages. It has a number of programmes on current affairs as well as on agriculture, rural development, health, family welfare, consumers rights and environment, etc. Doordarshan brings to the viewers all the major national and international sports and games through live telecasts. The entertainment programme includes music, dance, plays and serials.

TELETEXT SERVICE

Data transmission of information along with the picture in the invisible vertical blanking interval was introduced in Doordarshan in 1985. This service provides information on public facilities like timings of train and air service, stock market rates and weather forecasts, etc. This information can be retrieved by a decoder, which is required to be interfaced with the normal television receiver.

TELEVISION VIEWERSHIP

TV Audience

There are 65 million TV homes in the country. Smaller towns and villages account for three-fourth of all TV homes in the country. The audience share of Doordarshan in these homes is around 80 per cent. Table 8.6 provides the reach of TV vis-a-vis cable and other channels.

Table 8.6. Reach of TV and Cable Channels

Television homes in the country	65 million
Homes with cable connection	14 million
Reach of DD-1 (persons)	330 million
Reach of DD-2 (persons)	125 million
Reach of Satellite channels (persons)	70 million

Even in metro cities where the cable infiltration is large, Doordarshan has retained a substantial share of the audience as is evident from Fig. 8.8 below:

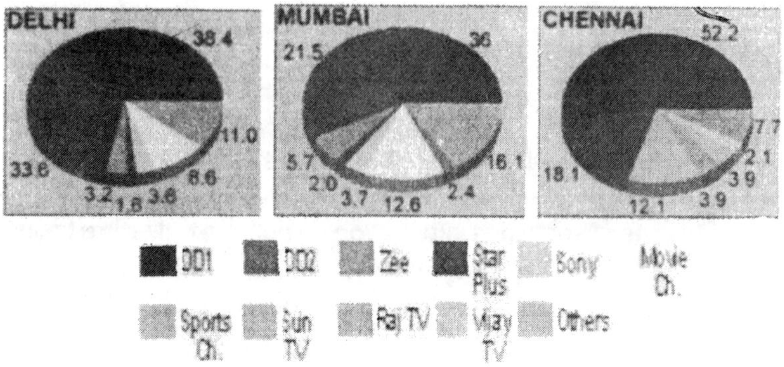

Fig. 8.8. Reach of Doordarshan vis-a-vis major satellite channels in Metros.
Source: IMRB People Meter Nov. 1997.

FUTURE BROADCASTING IN INDIA

The advent of digital technology and standardization of MPEG II[1] is leading towards convergence of telecommunications, computing and broadcasting. In the future, one can see television channels carrying more sound capabilities, mullet-channel surround sound, multi-lingual broadcasts, etc. DTH Service in Ku/ Ka band is already in the offing. With broadcasting going digital it will be able to use twisted pair cables to provide service using ADSL[2] technology. All India Radio and Doordarshan are planning to use the following types of Services.

Digital Transmission

Doordarshan has started digital transmission along with the analog transmissions, some of it's programmes with effect from 15th August 1999. The details are given in Table 8.7.

Table 8.7. Digital Transmission from Doordarshan.

			Down Link		
Channel	*Satellite*	*Analog MHz*	*Central Freq.MHz*	*Digital w.e.f. 15.08.1999 MPEG-2 DVB Compliant FEC.*	*Symbol Rate MS/Sec.*
DD-1	2E	3849	3830	¾	6
DD-2	2E	3929	3910	¾	6
News and Current Affairs	PAS-4	4155	4033	¾	6
DD	PAS-4	"	4033MPCC	¾	19.565
International		"	mode	"	"
DD Sports	PAS-4	"	"	"	

DIGITAL AUDIO BROADCASTING (DAB)

DAB is considered as the future of radio broadcasting. It can support 6 stereo/12 mono audio channels in a DAB multiplex of 1.5 MHz bandwidth. BBC has already started DAB service w.e.f. September 1995. AIR plans to introduce DAB service in its metro cities.

RADIO DATA SYSTEM

The frequency modulation service uses a large bandwidth to provide a programme service of high fidelity. It is possible to transmit various types of information on it in addition to the main programme and this additional information is not required to be related to that programme. Such application includes transmitting traffic information, the control of receivers, operation of display devices built into the receiver or operation of a separate display device such as a television receiver. Additionally it is possible to make use of the system for monitoring purposes. Taking advantage of this fact, a system called Radio Data System has been developed and it has been established that reliable reception of the supplementary information is possible within at least the same coverage area as provided for monophonic sound transmission. The supplementary information is digitally coded and amplitude modulated on a sub-carrier 57 kHz and multiplexed with the main carrier using an encoder. The receivers are required to be equipped with a built in decoder to retrieve the information at listener's will. AIR introduced Radio Paging Service in the year 1995 using this technology.

DIRECT TO HOME (DTH) SERVICE

Advent of the improved compression techniques allow more than one TV channel to be accommodated on a single TV transponder. With a small sized dish (Ku band DTH) of the order of 0.6 to 0.8 meter and an IRD (Integrated Receive Decoder), the viewer can receive a number of TV channels and select a particular channel of interest. The transmission being in digital mode, the sound and picture both are of extremely high quality. Due to better quality transmission, DTH will have a direct impact on broadcasting in the country. Doordarshan is awaiting enactment of legislation before starting this service

MULTI-POINT MULTI-CHANNEL DISTRIBUTION SYSTEM (MMDS)

Multi-point Multi-channel Distribution System (MMDS) is used for transmitting TV programmes which originate from a variety of sources as at the cable ends. All satellite delivered signals after they are processed to base-band formats are demodulated and converted to microwave frequencies of the MMDS. Terrestrial broadcast signals are passed through a heterodyne processor before being upconverted to the MMDS frequencies. MMDS uses omni-directional antenna having line of sight. The maximum range of a transmitting antenna can vary from 50 to 55 kms depending on the broadcast power, which may be in the range of 1 to 100 watts per channel. Typically a 10 watts transmitter would cover a radius of 25 kms. Due to it's very good picture quality, high reliability and ease of installation, Doordarshan is planning to introduce this service.

INTERNET BROADCASTING

Internet, the network of network, is fast becoming the global information Super Highway. It is now capable of delivering multimedia data including audio, video, text and graphics. While earlier it was not suited for audio and video delivery due to high compression requirements and long downloading time, the recent development of buffering, codec and streaming technologies have made the delivery of audio/video data using Internet. The buffering and streaming allow the data to be played immediately after receipt of few packets, without waiting for the full file to be downloaded. AIR ushered in the era of Internet audio broadcasting on 13.1.97 on experimental basis. The news and other data is uploaded daily on a server hired in USA[3]. This service allows user to receive the programme at a time convenient to him.

Doordarshan programmes have been made available on Internet w.e.f. 25.02.98 (website- http./dd india.net). The news and other data is uploaded daily on a server hired in USA. This service allows user to receive the programme at a time convenient to him.

CONCLUSION

The Indian Broadcasting network, its reach, programme content and status at the end of the twentieth century has been described in this chapter. A long journey has been made. However the journey in the next millennium will be more exciting and unpredictable. Broadcasting what we know today may totally be transformed in a new realm where it will encompass computers and telecommunication networks. The horizon is visible but hazy. Who can predict future? After all the eminent scientist Lord Kelvin stated in 1895 that "Heavier-than-air-flying machines are impossible" and in 1897 compounded this error by saying that "Radio has no future".

REFERENCES

1. MPEG Committee's International Standard, "Generic Coding of Moving Picture and Associated Audio", ISO-IEC Recommendation No. 13818, 9th November 1994.
2. General Introduction to Copper Access Technology by ADSL Forum at http://www.adsl.com/general_tutorial.html and ADSL Page at http://alumini.caltech.edu/ ~dank/isdn/adsl.html
3. http://www.air.kode.net

APPENDIX I

Colour and Broadcasting Systems by Country

Country	*TV*	*Colour*	*Stereo*	*Subtitles*
Albania	B/G	PAL		
Argentina	N	PAL-N		
Australia	B/G	PAL	FM-FM	TeleText
Austria	B/G	PAL	FM-FM	TeleText
Azores (Portugal)	B	PAL		
Bahamas	M	NTSC		
Bahrain	B	PAL		
Barbados	N	NTSC		
Belgium	B/G	PAL	Nicam	TeleText
Bermuda	M	NTSC		
Brazil	M	PAL M	MTS	
Bulgaria	D	SECAM		
Canada	M	NTSC	MTS	CC
Canary Is	B	PAL		
China	D	PAL		
Colombia	N	NTSC		
Cyprus	B	PAL		
Czechso-lvakia	D/K	SECAM/ PAL		
Denmark	B	PAL	Nicam	TeleText

Egypt	B	SECAM		
Faroe Islands (DK)	B	PAL		
Finland	B/G	PAL	Nicam	TeleText
France	E/L	SECAM		Antiope
Gambia	I	PAL		
Germany	B/G	PAL	FM-FM	TeleText
Germany (prev East)	B/G	SECAM/ PAL		
Gibraltar	B	PAL		
Greece	B/H	SECAM		
Hong Kong	I	PAL	Nicam	
Hungary	B/G & D/K	PAL (was SECAM)	Nicam (Budapest only)	
Iceland	B	PAL		
India	B	PAL		
Indonesia	B	PAL		
Iran	H	SECAM		
Ireland	I	PAL	Nicam	TeleText
Israel	B/G	PAL	Nicam	TeleText
Italy	B/G	PAL	FM/FM	TeleText
Jamaica	M	SECAM		
Japan	M	NTSC	Matrix	
Jordan	B	PAL		
Kenya	B	PAL		

Korea	M	NTSC		
Luxembourg	B/G	PAL	-	TeleText
Madeira	B	PAL		
Madagascar	B	SECAM		
Malaysia	B	PAL		
Malta	B/G	PAL		
Mauritius	B	SECAM		
Mexico	M	NTSC	MTS	CC
Monaco	L/G	SECAM/PAL		
Morocco	B	SECAM		
Netherlands	B/G	PAL	FM-FM	TeleText
New Zealand	B/G	PAL	Nicam	TeleText
North Korea	D/K?	SECAM		
Norway	B/G	PAL	Nicam	
Pakistan	B	PAL		
Paraguay	N	PAL		
Peru	M	NTSC		
Philipines	M	NTSC		
Poland	D/K	PAL		TeleText
Portugal	B/G	PAL	Nicam	TeleText
Romania	G	PAL		
Russia	D/K	SECAM		
Saudi Arabia	B	SECAM		
Seychelles	I	PAL		

Singapore	B	PAL		
South Africa	I	PAL		
South Korea	N	NTSC		
Spain	B/G	PAL	Nicam	TeleText
Sri Lanka	B/G	PAL		
Sweden	B/G	PAL	Nicam	TeleText
Switzerland	B/G	PAL	FM-FM	TeleText
Tahiti	KI	SECAM		
Taiwan	M	NTSC		
Thailand	B	PAL		
Trinidad	M	NTSC		
Tunisia	B	SECAM		
Turkey	B	PAL	-	TeleText
United Arab Emirates	B/G	PAL		
United Kingdom	I	PAL	Nicam	TeleText
Uruguay	N	PAL	MTS	
USA	M	NTSC	MTS	CC
Venezuela	M	NTSC		
Yugoslavia	B/H	PAL		
Zimbabwe	B	PAL		

APPENIXD II

Few Interesting Test Cards

Fig. 1. Test Card from Australia being used in 30 different counitries.

Fig. 2. Test Card from Sweden.

Fig. 3 NBC test pattern (1941)

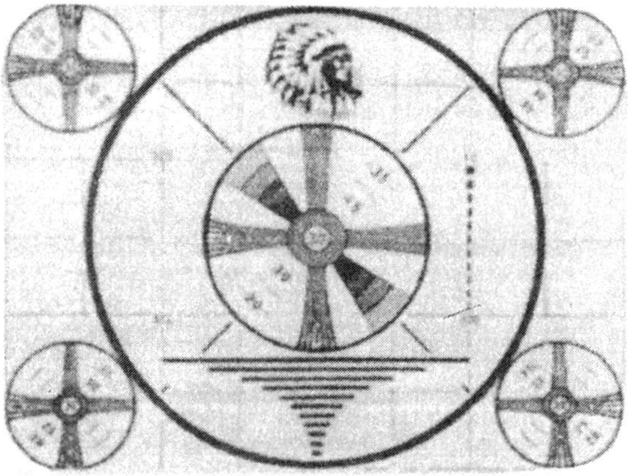

Fig. 4 RCA test pattern (1939)1939.

Fig. 5. 16:9 version of the Philips PM5644 test pattern

Fig. 6 Test pattern used by Telecom Italia Uplink in Fucino.

Fig. 7. Animated Colour Bars by France Telecom Uplink-Serte.

Fig. 8. Test image and caption from WTN London before they started scrambling their service.

Fig. 9 Test pattern from CBS

CREDITS, SOURCES AND REFERENCES

1. "History of Sound Recording and Broadcasting" at "http:// www.soundsite.com/ history/soundhis.html

2. Butterworth, E. William, *"Hi-Fi: From Edison's Phonograph to Quadrophonic Sound"* Four Winds Press, New York, 1977.

3. Hitchcock, Wiley, *"The Phonograph and Our Musical Life: Proceedings of a Centennial Conference"*, ed., Brooklyn College, Brooklyn, New York, 1977.

4. Jewell, Brian, Veteran, *"Talking Machines"*, Midas Books, Great Britain, 1977.

5. Maddox, Richard, *"The DAT Technical Service Handbook"*, Van Nostrand Reinhold, New York, 1994.

6. Read, Oliver and L. Walter Welch, *"From Tinfoil to Stereo: Evolution of the Phonograph"*, Howard W. Sams & Co. Inc., Indianapolis, 1976.

7. Basic Communication System at http://web.wse.nadn.navy.mil/ wse/academic/courses/es310/Lectures/AM.htm

8. Amplitude Modulation at http://www.antique-radio.org/terms/ amplifier.html

9. Ventura, Michael, "Hear That Long Snake Moan" essay from Shadow Dancing in the USA.

10. Television collection, (UK) National Museum of Film, Photography and Television at http://www.nmsi.ac.uk/nmpft/tv.htm

11. The Mechanical Era (of Television) at http://www.bravo.ca/mztv/ baird.html

12. The Unofficial Guide to the BBC: Television Menu at http:// www.vaxxine.com/master-control/BBC/bbc_tv.htm

13. John Logie Baird - Jones T&MM Encyclopaedia at http:// www.digitalcentury.com/encyclo/update/baird.html

14. The Media and Communications Study Site - (University of Wales,UK) at http://www.aber.ac.uk/~dgc/media.html
15. (Largely USA)Television / Media History Project at http://www.mediahistory.com/teevee.html
16. Royal Television Society NA, British TV Forum at http://www-royaltv.pp.asu.edu/british.html
17. Fonctionnement des disques lasers at http://www.artemis.jussieu.fr/dess/dossie94/groupe1/Laser/FONCTION.HTM
18. The Farnsworth Chronicles at http://www.edge.net/noma/philo/index.html
19. "The First Sparks - The Birth of Radio" at http://www.cinemedia.com.au/SFCV-RMIT-Annex/rnaughton/phd4410.html
20. "Just - 'The Cat's Whisker'- Magic Crystals" at http://www.cinemedia.com.au/SFCV-RMIT-Annex/rnaughton/phd4420.html
21. "Every Home Should Have One Radio for All" at http://www.cinemedia.com.au/SFCV-RMIT-Annex/rnaughton/phd4430.html
22. "Over and Out to You - Radio goes to War" at http://www.cinemedia.com.au/SFCV-RMIT-Annex/rnaughton/phd4440.html
23. "Pick A Box- The Golden Years of Radio" at http://www.cinemedia.com.au/SFCV-RMIT-Annex/rnaughton/phd4450.html
24. "Yeah,Yeah, Yeah - Long live the 'Trannie'" at http://www.cinemedia.com.au/SFCV-RMIT-Annex/rnaughton/phd4460.html">
25. "The Sound of Music - Good Guys, Glitter and Flares" at http://www.cinemedia.com.au/SFCV-RMIT-Annex/rnaughton/phd4470.html

26. "So much to listen to and so little time - A Walkman for the Train" at http://www.cinemedia.com.au/SFCV-RMIT-Annex/rnaughton/phd4480.html

27. "Surf's - Radio goes up on The Internet" at http://www.cinemedia.com.au/SFCV-RMIT-Annex/rnaughton/phd4490.html

28. "Hello, Angel... - What you want to hear, when you want to hear it." at http://www.cinemedia.com.au/SFCV-RMIT-Annex/rnaughton/phd4500.html

29. Ballou, Glen, *"Handbook for Sound Engineers"*, ed., The New Audio Cyclopedia. Indianapolis, IN: Howard W. Sams & Co., 1987.

30. Hartings, Robert L., *"Basic TV Technology"*, Boston, Focal Press, 1990.

31. Benson, K. Blair, *"Television Engineering Handbook"*, New York, McGraw-Hill, 1986.

32. Crutchfield, E.B., *"National Association of Broadcasters Engineering Handbook"*, ed. 7th ed. Washington, DC, NAB, 1985.

33. Noll, A. Michael, *"Television Technology: Fundamentals and Future Prospects"*, Norwood, MA, Artech, 1988.

34. Rzeszewski, Theodore S., *"Television Technology Today"*, ed. New York, Institute of Electrical and Electronic Engineers, 1985.

35. Cartwright, Steve R., *"Training With Video"*, White Plains, NY, Knowledge Industry Publications, 1986.

36. Gayeski, Diane, *"Corporate and Instructional Video"*, 2nd ed. Englewood Cliffs, N.J., Prentice-Hall, 1991.

37. Utz, Peter, *"Video User's Handbook"*, 3rd ed. New York, Prentice-Hall, 1989.

38. Prentiss, Stan., *"HDTV/DTV: High Definition Television"*, Blue Ridge Summit, PA, TAB Books, 1990.

39. Pensinger, Glen, *"4:2:2 Digital Video: Background and Implementation"*, ed. White Plains, NY, Society of Motion Picture and Television Engineers, 1989.

40. Prentiss, Stan., *"Television from Analog to Digital"*, Blue Ridge Summit, PA, TAB Books, 1985.

41. McWhorter, Gene (Eugene W.), *"Understanding Digital Electronics"*, 2nd ed. Dallas, TX, Texas Instruments, 1984.

42. Medoff, Norman J. and Tanquary, Tom, *"Portable Video ENG and EFP"*, White Plains, NY, Knowledge Industry Publications, 1986.

43. Camras, Marvi, *"Magnetic Recording Handbook"*, New York, Van Nostrand Reinhold, 1988.

44. Jorgensen, Finn, *"The Complete Handbook of Magnetic Recording"*, 3rd ed. Blue Ridge Summit, PA, TAB Books, 1988.

45. Clifford, Martin, *"Microphones"*, 3rd ed. Blue Ridge Summit, PA, TAB Books, 1986.

46. Huber, David Miles, *"Microphone Manual: Design and Application"*. Indianapolis, IN, Howard W. Sams & Co., 1988.

47. Jarvis, Peter, *" A Production Handbook: A Guide to the Pitfalls of Programme Making"*, Newton, MA, 1993.

48. Alkin, Glyn. *"Sound Recording and Reproduction"*, Newton, MA, Focal Press, 1991.

49. Rumsey, Francis, *"Sound and Recording: An Introduction"*, 2nd ed. Newton, MA, Focal Press, 1994.

50. Bartlett, Bruce, *"Introduction to Professional Recording Techniques"*, Indianapolis, IN, Howard W. Sams & Co. 1987.

51. Runstein, Robert E. and David Miles Huber, *"Modern Recording Techniques"*, 2nd ed. Indianapolis, Howard W. Sams & Co., 1986.

52. Ford, Ty., *"Advanced Audio Production Techniques"*, Newton, MA, Focal Press, 1993.

53. Gross, Lynne, and David E. Reese, *"Radio Production Worktext: Studio and Equipment"*, Boston, Focal Press, 1990.

54. O' Donnell, Lewis B., *"Modern Radio Production"*, 2nd ed. Belmont, Calif. Wadsworth Publishing Co., 1990.

55. Alkin, Glyn, *"Sound Techniques for Video and TV"*, 2nd ed. Boston, Focal Press, 1989.

56. Huber, David Miles, *"Audio Production Techniques for Video"*, Indianapolis, Howard W. Sams, 1987.

57. "Multichannel TV Sound: How Audio Came to Video", Washington, DC, *Television Digest*, Inc., 1985.

58. Alten, Stanley R., *"Audio in Media"*, 3rd ed. Belmont, CA, Wadsworth, 1990.

59. Nisbett, Alec, *"The Sound Studio"*, 5th ed. Newton, MA, Focal Press, 1993.

60. Oringel, Robert S., *"Audio Control Handbook: For Radio, and Television Broadcasting"*, 6th ed. Boston, Focal Press, 1989.

61. Thom, Randy, *"Audio Craft: An Introduction to the Tools and Techniques of Audio Production"*, 2nd ed. Washington, DC, National Federation of Community Broadcasters, 1989.

62. Fielding, Ken, *"Introduction to Television Production"*, New York, Longman, 1990.

63. Whittaker, Ron, *"Television Production. Mountain View"*, CA, Mayfield Publishing Co., 1993.

64. Wurtzel, Alan and Stephen R.Acker, *"Television Production"*, 3rd ed. New York, McGraw-Hill, 1989.

65. Wiegand, Ingrid, *"Professional Video Production"*, White Plains, NY, Knowledge Industry Publications, 1985.

66. Compesi, Ronald J. and Ronald E. Sherriffs, *"Video Field Production and Editing"*, Boston, Allyn and Bacon, 1994.

67. DiZazzo, Raymond, *"Corporate Television: A Producer's Handbook"*, Boston, Focal Press, 1990.

68. Millerson, Gerald, *" Video Production Handbook"*, 2nd ed. Boston, Focal Press, 1992.

69. Utz, Peter, *"Today's Video: Equipment, Setup, and Production"*, Englewood Cliffs, NJ, Prentice-Hall, 1987.

70. Rowlands, Avril, *"Continuity in Film and Video"*, 2nd ed. Boston, Focal Press, 1989.

71. Musburger, Robert B., *"Single Camera Video Production"*, Newton, MA, Focal Press, 1992.

72. Rowlands, Avril, *"The Continuity Handbook: for Single Camera Shooting"*, Newton, MA, Focal Press, 1994.

73. Wilkie, Bernard, *"The Technique of Special Effects in Television"*, 2nd ed. Boston, Focal Press, 1989.

74. Dancyger, Ken, *"The Technique of Film and Video Editing"*, Newton, MA, Focal Press,1993.

75. Kerner, Marvin M., *"The Art of the Sound Effects"*, Editor. Boston, Focal Press, 1989.

76. Mott, Robert L., *"Sound Effects: Radio, TV and Film"*, Boston, Focal Press, 1990.

77. Anderson, Gary H., *"Video Editing and Post-Production: A Professional Guide"*, 2nd ed. White Plains, NY, Knowledge Industry Publications, 1988.

78. Browne, Steven E., *"Videotape Editing: A Post-production Primer "*, 2nd ed. Boston, Focal Press, 1993.

79. Anderson, Gary H., *"Electronic Post-Production: The Film to Video Guide"*, White Plains, NY, Knowledge Industry Publications, 1986.

80. Ohanian, Thomas D., *"Digital Nonzlinear Editing"*, Newton, MA, Focal Press, 1993.

81. Schneider, Arthur, *"Electronic Post Production: Terms and Concepts"*, Newton, MA, Focal Press, 1990.

82. Schneider, Arthur, *"Electronic Post-production and Videotape Editing"*, Boston, Focal Press, 1989.

83. Pohlmann, Ken C., *"Principles of Digital Audio"*, 2nd ed. Indianapolis, IN, Howard W. Sams & Co., 1989.

84. Watkinson, John, *"The Art of Digital Audio"*, 2nd ed. Newton, MA, Focal Press, 1994.

85. Watkinson, John, *"R-DAT"*, Newton, MA, Focal Press, 1991.

86. Baert, L., et. al., *"Digital Audio and Compact Disc Technology"*, 2nd ed. Newton, MA, Focal Press, 1992.

87. Sweeney, Daniel, *"Demystifying Compact Discs: A Guide to Digital Audio"*, Blue Ridge Summit, PA, TAB Books, 1986.

88. Lambert, Steve and Jane Sallis, *"CD-1 and Interactive Videodisc Technology"*, eds. Indianapolis, IN, Howard W. Sams & Co., 1986.

89. Souter, Gerald A., *"The Disconnection: How to Interface Computers and Video"*, White Plains, NY, Knowledge Industry Publications, 1988.

90. Stokes, Judith Tereno, *"Microcomputers in TV Studios"*, White Plains, NY, Knowledge Industry Publications, 1986.

91. Watkinson, John, *"An Introduction to Digital Video"*, Newton, MA, Focal Press, 1994.

92. Watkinson, John, *"The Art of Digital Video"*, 2nd ed. Newton, MA, Focal Press, 1993.

93. Watkinson, John', *"The Digital Videotape Recorder"*, Newton, MA, Focal Press, 1994.

94. Prentiss, Stan., *" AM Stereo and TV Stereo: New Sound Dimensions"*, Blue Ridge Summit, PA, TAB Books, 1985.

95. Rumsey, Francis, *"Stereo Sound for Television"*, London, Boston, Focal Press, 1989.

96. Kallenberger, Richard H. and George D. Cvjetnicanin, *"Film Into Video: A Guide to Merging the Technologies"*, Newton, MA, Focal Press, 1994.

97. Kuney, Jack, *"Take One: Television Directors on Directing"*, Westport, CT, Greenwood/Praeger, 1990.

98. Lewis, Colby, and Tom Green, *" The TV Director/Interpreter"*, Rev. ed. New York, Hastings House, 1990.

99. Rabiger, Michael, *" Directing the Documentary"*, Boston, Focal Press, 1987.

100. Verna, Tony, *"Live TV: An Inside Look at Directing and Producing"*, Boston, Focal Press, 1987.

101. Kehoe, Vincent J.R., *"The Technique of the Professional Make-up Artist for Film, Television, and Stage"*, Boston, Focal Press, 1985.

102. Rowlands, Avril, *"The Production Assistant in TV and Video"*, Boston, Focal Press, 1987.

GLOSSARY

A/A Roll Editing

Editing from a single source VCR using a freeze frame as a transition from the source VCR to itself (source "A" to "A").

A/B Roll Editing

Editing from two VCRs to a third recording VCR. A switcher or mixer is used to provide effects such as dissolves.

AGC (Automatic Gain Control)

A circuit that automatically adjusts audio or video input levels.

Aliasing

Undesirable video display effects caused by excessive high frequency video information. For example, Jaggies or Stair-stepping—Stepped or jagged edges of angled lines, especially at the slanted edges of letters. There is Raster scan aliasing, e.g., twinkling or strobing effects on sharp horizontal lines and Temporal aliasing, e.g., rotating wagon wheel spokes apparently reversing direction.

Analog

An electrical signal with continuously varying electrical voltages. Analog video after editing for several generations suffers from generation loss and is subject to degradation due to noise and distortion.

Animatic

Simple animation consisting of art work designed to be used as a videotape storyboard.

Anti-Aliasing

The process of electronically reducing aliasing, especially letters and genlocked graphic elements.

AM (Amplitude Modulation)

Amplitude modulation is a process used for some radio (AM broadcast) and television video transmission. A low frequency (program) signal modulates (changes) the amplitude of a high frequency RF carrier signal (causing it to deviate from its nominal base amplitude). The original program signal is recovered (demodulated) at the receiver. This system is extensively used in broadcast radio transmission because it is less prone to signal interference and retains most of the original signal quality.

Aperture

An adjustable opening in a lens which, like the iris in the human eye, controls the amount of light entering a camera. The size of the aperture is controlled by the iris adjustment and is measured in f-stops. A smaller f-stop number corresponds to a larger opening causing more light to pass. F-stop examples are F2, F2.8, F4, F5.6, F8, F11. F-stops are logarithmic. Each stop admits 100 per cent more light than the previous one.

Aspect Ratio

Ratio of picture height to picture width in video and TV systems is called aspect ratio.

Assembly Edit

An edit wherein all existing signals on a tape are replaced with new signals. Assembly edits cannot be used for editing because they erase the control track portion of the videotape.

Auto Assembly

Process of assembling an edited videotape on a computerized editing system using an edit decision list.

B Roll

Cutaway shots which are used to cover the visual part of an interview or narration. The term is often used in TV news.

Betacam

The most common broadcast quality video format. Betacam SP is the enhanced version.

Betamax

The obsolete home video format. Lost the format battle to VHS even though it was slightly superior. The cassette size, however, went on to become BETACAM.

Black Burst

A composite colour video signal comprised of sync, colour burst and black video. Used to synchronize (genlock) other video sources to the same sync and colour information. Black burst generators are used in editing systems "lock" the entire facility to a common signal ("house sync" or "house black").

Black Level

Voltage in a video signal corresponding to black.

Blanking Level

Also known as the pedestal, it is the voltage level produced at the end of each horizontal picture line that separates the portion of the video signal containing the picture information from the portion containing the synchronizing information. This voltage makes the electron beam "invisible" as it moves to draw the next visible line.

Blandking Interval (Horizontal and Vertical)

The horizontal blanking interval is the time between the end of one scan line and the beginning of the next. The vertical blanking interval is the time between the end of one video field and the beginning of the next. Blanking occurs when a monitor's electron beam is positioned to start a new line or a new field. The blanking interval is used to instantly reduce the beam's amplitude so that the return trace is invisible. The screen goes blank for a fraction of second. (See Vertical Intercal Switching)

BNC Connector

A type of professional connector used on some VCRs, cameras and video equipment providing twist-lock capability.

Boom

An overhead pole device used to position a microphone close to the actors, but out of the shot. A fishpole is the portable version.

CCD

Charged Coupled Device is an integrated circuit used for capturing video images. It has largely replaced tubes in modern video cameras.

CCTV (Closed Circuit TV)

A video system used in commercial internal installations for security, medical and educational.

CU

Close-up shot.

CG (Character Generator)

An electronic typewriter used to create titles for video.

CATV

Acronym for cable TV, derived from the older term, community antenna television.

Chroma

The colour information in a video signal, consisting of hue (phase angle) and saturation (amplitude) of the colour sub-carrier signal.

Chroma Corrector

A device used to correct problems related to the chroma of the video signal, as well as colour balance and colour noise.

Chroma Noise

Noise, which manifests itself in a video picture as coloured snow.

Chroma Key

The process of overlaying one video signal over another by replacing a range of colors with the second signal. Typically, the first (foreground) picture is photographed with a person or object against a special, single-color background (the key-color). The second picture is inserted in place of the key-color. The most common example is in broadcast weather segments where pictures of weather maps are inserted "behind" the talent.

Chrominance and Chrominance Level

The color portion of a video signal separated from the luminance component, representing the saturation and tint at a particular point of the image. Black, gray and white have no chrominance, but any colored signal has both chrominance and luminance. The higher the chrominance level, the stronger the color (e.g., a strong signal produces red, and a weak signal, pink).

Chromakey

Electronically matting or inserting an image from one camera into the picture produced by another. Also called "keying." The subject to be inserted is shot against a solid primary color background. Signals from the two sources are merged through a special effects generator.

Chrominance

The color portion of a video signal is called chrominance.

Clipping

The electronic process of cutting off the peaks of either the white or black excursions of a video signal to limit the signal. Sometimes, clipping is performed prior to modulation, and sometimes to limit the signal, so it does not exceed the limits of the composite video signal (7.5 and 100 IRE units).

Color Bars

A standard video test pattern, which includes samples of primary and secondary colors.

Color Burst

The portion of a color video signal, which contains a short sample of the color sub-carrier used to add color to a signal. It is used as a color synchronization signal to establish a reference for the color information following it and is used by a color monitor to decode the color portion of a video signal. The color burst acts as both amplitude and phase reference for color hue and intensity. The color oscillator of a color television receiver is phase locked to the color burst.

Color Correction

A process in which the coloring in a television image is altered or corrected by electronic means.

Color Decoder

A device, which divides a video signal into its basic color components. In TV and video, color decoding is used to derive signals required by a video monitor from the composite or Y/C signals.

Color Phase

The phase of the chroma signal as compared to the color burst, is one of the factors that determines a video signal's color balance.

Color Temperature

A method for measuring the overall color of a light source, measured in degree Kelvin. Sunny daylight is approximately 5500 K. Overcast daylight is higher. Fluorescent lights are approx. 4100 K. Indoor incandescent lights are 2800 K and professional movie lights are 3200 K. Higher numbers indicate bluer light, lower numbers indicate a warmer light. The color temperature of the lighting must match the color temperature of the camera. In video this is accomplished by setting the white balance of the camera.

Color Sub-Carrier

The carrier frequency (3.58 MHz in NTSC and 4.43 MHz in PAL) on which the color information is provided. Color TV sets use special circuits which decode the color component for accurate display.

Component Video

Video signal in which luminance and synch information are re-corded separately from the color information. Formats such as Betacam, SVHS and Hi-8 use component signals. Component video comes in several forms: RGB (red, green, blue), YUV (luminance, sync, and red/blue) and Y/C (luminance and chrominance). Y/C is also called S-Video used in the S-VHS and Hi-8 formats.

Composite Video

A video signal formats such as VHS, in which the luminance and chrominance elements have been combined.

Composite Sync

A signal consisting of horizontal sync pulses, vertical sync pulses and equalizing pulses only.

Compression

The process of electronically processing video signals so that it requires less storage on a computer hard drive.

Contrast

The degree to which luminance values contain very dark and very light values. A high-contrast picture has more black and white values with fewer values in between. A low contrast pic-ture has more middle tones without very dark or very light areas.

Control-L

Sony's editing control protocol, which allows two-way communication between a camcorder or VCR and an edit controller.

Control-M

Panasonic 5-pin edit control protocol.

ControlL-S

Sony transport control protocol which duplicates a consumer VCR's infra-red remote transport control. Unlike Control-L, Control-S does not allow the controller to read tape counter information.

Controal Track

Type of video editing that controls the in and out points of edits by counting pulses on a control track portion of the videotape. The pulses are counted by the edit controller to perform fairly accurate editing. Edit controllers that read time-code make more accurate edits.

Conforming

On-line editing to create the final edit master. The off-line edit master is used as a guide.

Continuity

Controlling the elements in a shot to insure that edits will flow smoothly and produce a coherent motion picture story without jarring the viewer.

Crawl

Text that moves horizontally across the screen is called crawl.

Crossfade

The audio equivalent of the video picture dissolve. The first sound track gradually fades out while the second sound track simultaneously replaces it.

Crosstalk

The interference between two audio or two video signals. In audio crosstalk this signal leakage may occur between the left and right channels. In video, crosstalk between channels can be luminance/sync crosstalk or chroma crosstalk. Video crosstalk can cause ghost images from one source appear over the other.

Cue Card

A card with the actor's lines written on it to enable the actor to read or remember his lines.

D1/D2/D3

Professional digital video formats. The D1 system uses component video. The D2 and D3 systems use composite video. There is no D4 format. Digital formats do not suffer from the generation loss inherent in analog formats.

DVE

Digital Video Effects. A shot can bend, twist and fold into various shapes.

DVI (Digital Video Interface)

Multimedia standard for computer generated text and graphics, which can be transferred to video.

DAT (Digital Audio Tape)

An audio recording and playback format developed by Sony, with a signal quality capability surpassing that of the CD.

dB (Decibel)

A logarithmic unit, which expresses the ratio between two amounts of electric or acoustic signal power. Used for measuring the strength of audio and video signals.

Decode

To separate a composite video signal into its component elements is called decoding.

Delay Correction

When an electronic signal travels through electronic circuitry or long cable runs, delay problems may occur. This causes a displaced image. Special circuits are used to correct the delay

Demodulator

An electronic circuit, which separates the audio and video signals from the RF carrier frequency.

Depth of Field

The range of objects in front of a camera lens, which are in focus. Smaller f-stops provide greater depth of field.

Digital

A system whereby a variable analog signal is broken down and encoded into discrete binary bits of ones and zeros. These numbers represent a mathematical model of the original signal. When copied, they do not degrade as an analog signal does. An analog-to-digital (A/D) converter chip takes samples of the signal at a fixed time interval known as sampling frequency. This digital stream can be recorded onto magnetic media. Upon playback, a digital-to-analog (D/A) converter chip reads the binary digital signals which are virtually immune to noise, distortion, crosstalk, and other quality problems.

DIN (Deutsche Industrie Norme)

An international connector standard. DIN connectors carry both audio and video signals and are common equipment in Europe.

Dissolve

A video or film transition where one shot gradually fades out while a second shot fades in.

Distribution Amplifier

A device which, splits and amplifies an audio and/or video source tape or signal to several audio/video outputs. Used to duplicate one videotape to any number of VCRs with minimal loss of signal strength.

Dolby (TM)

A compression/expansion (companding) noise reduction system developed by Ray Dolby, widely used in consumer, professional and broadcast audio applications. Signal-to-noise ratio improvement is accomplished by processing a signal before recording and reverse-processing the signal upon playback.

Dolly

Dolly is a camera platform on wheels. To dolly is to smoothly bring the camera closer or farther from the subject.

Downstream Keying

Electronically superimposing text or graphics over a scene (luminance key) or of placing one video image into another (chroma key). The Downstream Key signal must be genlocked to the other signals.

Dropout

A defect on the videotape, which causes a brief flash of a horizontal black line on the screen. Commonly found at the beginning and end of tapes. The quality of videotape is graded by the number of dropouts and priced accordingly.

Dub

Duplicate copy of a videotape.

EIA RS-170A

The timing specification standard for NTSC broadcast video equipment.

Encode

The process of combining analog or digital video signals, e.g., red, green and blue, into a composite signal.

Encode

A circuit that combines the primary red, green and blue signals into a composite video signal.

ECU

Extreme close-up shot.

EDL (Edit Decision List)

A complete list of time-code numbers for each shot and sound used in the offline edit master. These time-code numbers are used to create the final online edit master.

Establishing Shot

A wide shot, showing much of the location.

Edit

Edit is the process or result of selectively recording video and/or audio on finished videotape. Typically involves reviewing raw footage and transferring desired segments from master tape(s) onto new tape in a pre-determined sequence.

Edit Controller

Electronic programmer used in conjunction with VCRs/ camcorders to facilitate automated videotape editing with speed, precision, and convenience.

Edit Decision List (EDL)

Handwritten or computer-generated compilation of all post-production edits to be executed in a video work is called EDL.

EFP (Electronic Field Production)

Film-style production approach using a single camera to record on location. Typically shot for post-production application, non-live feed.

8 mm

Compact videocassette format, popularized by camcorders, employing 8 millimeter wide videotape.

Electret Condenser

Microphone type incorporating a precharged element, eliminating need for bulky power sources.

Encode

The process of combining analog or digital video signals, e.g., red, green and blue, into a composite signal.

Encoder

"Combiner" device that translates a video signal into a different format viz. RGB to composite. Horizontal and vertical sync information joins individual red/green/blue components.

ENG (Electronic News Gathering)

Use of portable video cameras, lighting and sound equipment to record news events in the field quickly, conveniently, and efficiently.

Equalization

Emphasizing specific audio or video frequencies and eliminating others as signal control measure, usually to produce particular sonic qualities which are achieved with equalizer.

Essential Area

Boundaries within which contents of a television picture are sure to be seen, regardless of size differences in receiver displays. Also called "critical area" and "safe title area," encompasses 80 per cent of total screen.

Establishing Shot

Opening picture of a program or scene. Usually a wide and/or distant perspective, orients viewer to overall setting and surroundings.

Fade

Gradual diminishing or heightening of visual and/or audio

intensity is called fading. "Fade out" or "fade to black," "fade in" or "up from black" are common terms.

FCC

(Federal Communications Commission) The U.S. Agency, which governs radio and television broadcasting.

Feed

Act or result of transmitting a video signal from one point to another.

Feedback

Infinite loop of visual patterns from signal output being fed back as input; achieved by aiming live camera at receiving monitor.

Field

One-half of a television frame, containing all the odd or even scanning lines of the picture. In NTSC it is 262.5 horizontal lines at 59.94 Hz. In PAL, it is 312.5 lines at 50 Hz.

Field of View

Extent of a shot that is visible through a particular lens is called field of view.

Fill Light

Supplementary illumination, usually from a floodlight positioned mid-way between camera and subject, which lightens or eliminates shadows created by key light.

Film Chain

Projectors, multiplexors and cameras, used to transfer film to video.

Filter

Transparent material, typically glass accessory, mounted at front of camcorder lens to regulate light passing through. Manipulates colors and image patterns, often for special effect purposes.

Flare

Bright flashes and/or extreme contrast reduction evident in picture, caused by excessive light beaming into a camera's lens and reflecting off its internal glass elements.

Flat Lighting

Illumination characterized by even, diffused light without shadows, highlights, or contrast. May impede viewer's sense of depth, dimension, drama.

Floodlight

Radiates a diffused, scattered blanket of light with soft, indistinct shadows. Best used to spread illumination on broad areas, whereas spotlights focus on individual subjects.

Fluid Head

Tripod mount type containing viscous fluid, which lubricates moving parts, dampens friction. Design facilitates smooth camera moves, alleviates jerkiness.

Focal Length

A measurement of the magnification of a lens indicated in millimeters. A zoom lens has a variable focal length, which allows the camera to film closer or farther from the subject without moving the camera or subject.

Foley

Personal sound effects, like footsteps, breathing or punches used to heighten realism.

Follow Focus

Controlling lens focus so that an image maintains sharpness and clarity despite camera and/or subject movement.

Format

Videotape and video equipment design differences—physical and technical—dictating compatibility and quality. In most basic sense, refers to standardized tape widths, videocassette sizes.

Frame

The smallest increment of a complete television picture, equal to one-thirtieth of a second is called a frame.

Framestore

A digital device designed to store and display a single frame of video as a freeze frame.

Frame Synchronizer

Digital device, which synchronizes two or more video signals. The frame synchronizer uses one of its inputs as a reference and genlocks the other video signals to the reference sync and color burst signals. By delaying the other signals so that each line and field starts at the same time, two or more video images can be blended, wiped and otherwise processed together.

Frame Grabber

High-speed digitizer capable of capturing frames at rate necessary to create real-time motion.

Framing

Act of composing a shot in the camcorder's viewfinder for desired content, angle, and field of view—overall composition.

Freeze Frame

Single frame paused and displayed for an extended period during video playback; suspended motion perceived as still snapshot.

Frequency

Number of vibrations produced by a signal or sound, usually expressed as cycles per second, or hertz (Hz).

Frequency Response

Measure of the range of frequencies a medium can respond to and reproduce. Good video response maintains picture detail; good audio response accommodates the broadest range, most exacting sound.

Friction Head

Tripod mount type with strong spring that counterbalances camera weight, relying on friction to hold its position. More appropriate for still photography than movement-oriented videomaking.

F-stop

Numbers corresponding to variable size of camera's iris opening, and thus amount of light passing through lens. The higher the number, the less light enters.

Full-motion Video

A standard for video playback on a computer; refers to smooth-flowing, full-color video, similar to a VCR or television. Also known as full-screen, full- or true-color, or full-motion video.

Gain

Video amplification, signal strength. "Riding gain" means varying controls to achieve desired contrast levels.

Generation

Relationship between a master video recording and a given copy of that master. A copy of the original master constitutes a second-generation duplication.

Generation Loss

Created when editing or copying one analog videotape to another videotape. Most apparent in less expensive video formats. Theoretically absent from digital video editing.

Genlock/Encoder

Device, which allows computer text and graphics to be recorded or superimposed on video. Also includes a fader to fade video or computer graphics.

Genlock

A system whereby the internal sync generator in a device such as a camera, locks on to and synchronizes itself with a separate incoming signal.

Ghosting

Undesirable faint double screen image caused by signal reflection or improperly balanced video circuitry. "Ringing" appears as repeated image edges.

Glitch

Momentary picture disturbance.

Grain

Blanketed signal noise viewed as fuzziness, unsmooth images—attributable to luminance inadequacies.

Hard Disks

Common digital storage component in a computer. For video use, hard disks need: (1) an access time of less than 10 milliseconds; (2) a sustained through put (data transfer rate) of 3 Megabytes per second; and (3) a maximum time for housekeeping of 33 milliseconds (one video frame).

Hard Light

Type of light that creates brilliant highlights and sharp shadows.

HDTV (High-Definition Television)

Television system standard affording greater resolution for sharper pictures and wide-screen viewing via specially-designed TV equipment.

Head

Electromagnetic components within camcorders and VCRs that record, receive, and erase video and audio signals on magnetic tape. [2] Tripod's camera mount.

Headroom

Space remaining between the top of a subject's head and a monitor's upper screen edge.

Hi 8

Improved version of 8 mm videotape format characterized by higher luminance resolution for a sharper picture. Compact "conceptual equivalent" of Super-VHS.

Hi-Fi

(High Fidelity) Generalized term defining audio quality approaching the limits of human hearing, pertinent to high-quality sound reproduction systems.

Hiss

Primary background signal interference in audio recording, result of circuit noise from a playback recorder's amplifiers or from a tape's residual magnetism.

Horizontal Resolution

Specification denoting amount of discernable detail across a screen's width.

Image Enhancer

Video signal processor that compensates for picture detail losses and distortion occurring in recording and playback. Exaggerates transitions between light and dark areas by enhancing high frequency region of video spectrum.

Insert Edit

An electronic edit in which the existing control track is not replaced during the editing process. The new segment is inserted onto a pre-recorded black videotape.

Insert Shot

A close-up shot used to hide an edit or to emphasize a detail.

Interlace

The manner in which a television picture is composed scanning alternate lines to produce one field, approximately every 1/60 of a second in NTSC. Two fields comprise one television frame resulting in the NTSC television frame rate of approximately 30 fps.

Iris

Camcorder's diaphragm lens opening or aperture that regulates amount of light entering camera. Size is measured in f-stops.

ISO

International Standards Organization.

Jitter

Video image aberration seen as slight, fast vertical or horizontal shifting of a picture or portion of one.

Jump Cut

Unnatural, abrupt switch from and to shots identical in subject but slightly different in screen location. Awkward progression makes subject appear to jump from one screen location to another. Remedied with cutaway. A jarring edit caused by the choice of shots rather than any technical imperfection.

Kelvin

Temperature scale used to define the color of a light source; abbreviated as "K".

Key Light

Principal illumination source on a subject or scene, normally positioned slightly off centre and angled to provide shadow detail.

Laser

An acronym for Light Amplification by Stimulated Emission of Radiation.

Lavaliere

Small, easily concealed, unobtrusive, and aesthetically pleasing microphone, typically attached to clothing or worn around the neck for interview settings.

Linear Editing

Analog, tape-based editing, called linear because scenes are laid in a line along the tape.

Lip Sync

Proper synchronization of video with audio—lip movement with audible speech. Better known as technique widely practiced with music video recordings, whereby "vocalists" mime to playback of pre-recorded music.

Location

Any place filming occurs except a studio.

Log

A paper listing of the time-code addresses of shots, scenes and takes. The log is an efficient way to find shots during editing.

Long Shot

Camera view of a subject or scene, usually from a distance, showing a broad perspective.

LPX (Long Play)

Middle tape speed of a VHS VCR, accommodating four-hour recordings.

LTC (Longitudinal Time Code)

Frame identification numbers encoded as an audio signal and recorded lengthwise on the edge of a tape.

Luminance

Black and white portion of video signal, carries brightness information representing picture contrast, light and dark qualities; frequently abbreviated as "Y".

Lux

Amount of lumens in a square meter. Means of measuring a camcorder's low-light sensitivity—minimum amount of illumination required to record an "acceptable" image. Lower the lux reading the greater the sensitivity.

Macro

Lens capable of extreme close-up focusing, useful for intimate views of small subjects.

Master

Original recorded videotape footage; "edited master" implies original copy of tape in its edited form.

Medium Shot

Defines any camera perspective between long shot and close-up, whereby subjects are viewed from medium distance.

Mic

"Mike," short for "microphone".

Mix

Combining two or more sound sources, with various channels controlled to achieve desired balance of single audio signal output. Executed with audio mixer. Combining video signals from two or more sources.

Monitor

A video display similar to a TV, but having superior visual quality and without a tuner. An audio monitor is a speaker.

Montage

Rapid sequence of video shots assembled to communicate a particular image or mood. Juxtaposition of seemingly unrelated material can conjure new idea or message.

MPEG 1

Refers to a standard method of compressing audio and video data down to 1/50th of their original size. This standard was developed by the Motion Pictures Experts Group.

Multiplexer

Device for mixing television signals to a single video recorder.

Natural Light

Planetary illumination—from sun, moon, stars—be it indoors or outdoors. Has higher color temperature than artificial light, and thus more bluish qualities.

NiCad

(Nickel Cadmium) Abbreviation coined and popularized by SAFT America for lightweight camcorder battery type designed to maintain power longer than traditional lead-acid batteries.

Noise

Undesirable video or audio signal interference typically seen as snow, heard as hiss.

Non-linear Editing

Digital "cut and paste" editing that uses a hard drive instead of tape to store images. Random access allows easy arrangement of scenes in any order. Also eliminates the need for rewinding and allows for multiple dubs without generation loss.

Offline

The creative editing process which uses copies of the camera tapes on a typically "cuts only" inexpensive editing system. All creative decisions and approvals are made during this process.

Omnidirectional

Microphone pick-up pattern whereby sound is absorbed equally from all directions.

Online

The final technical editing process, which uses the original camera tapes to repeat all decisions made in the offline editing process. Online editing uses a more sophisticated and expensive editing system capable of transitions like dissolves and wipes.

PAL (Phase Alternate Line)

(Phase Alternate Line) 625 line 25 frame per second television signal standard.

Pan

Horizontal camera pivot, right to left or left to right, from a stationary position. Follows a subject, re-directs viewer's attention from one subject to another, shows relationships between subjects, and scans subjects too large to fit into one shot.

Paper Edit

A list of edits made entirely on paper by viewing Window Dub copies of the original camera tapes.

Pick-up

[1] A video camera's image sensing element, either CCD (charge coupled device) or MOS (metal oxide semiconductor); converts light to electrical energy. [2] A microphone's sound reception.

Pickup Pattern

Defines a microphone's response to sounds arriving from various directions or angles.

PIP

(picture in picture, p-in-p, pix in pix) Image from a second video source inset on a screen's main picture, the big and small pictures usually being interchangeable.

Playback

Videotaped material viewed and heard as recorded, facilitated by camcorder or VCR.

Post-production

Any video production activity following initial recording. Typically involves editing, addition of background music, voiceover, sound effects, titles, and/or various electronic visual effects. Results in completed production.

Posterization

Electronic special effect transforming a normal video image into a collage of flattened single-colored areas, without graduations of color and brightness.

Pre-production

The vital phase of production in which the script, budget, locations, actors and props are planned.

Pre-roll

[1] Slight backing-up function of camcorders and VCRs when preparing for tape recording; ensures smooth, uninterrupted transitions between scenes. [2] Usually for on-air applications, starting tape playback earlier than necessary to ensure full operating speed and stabilization.

Proc Amp (Processing Amplifier)

(Processing Amplifier) Video image processor that boosts video signal's luminance, chroma, and sync components to correct such problems as low light, weak color, or wrong tint.

Production

The actual filming and creation of the raw elements as required by the script.

Props

Short for "properties," objects used either in decorating a set (set props) or by talent (hand props).

PCM

Pulse Code Modulation.

Raster

The area of a TV picture tube that is scanned by the electron beam. Also the active area of visual display on a TV, monitor or any cathode ray tube (CRT).

Raw footage

Pre-edited recordings, usually direct from camera.

Real Time

The actual time during which video recording occurs, distinguished from the tampering of time via editing.

Receiver

Television set that includes a tuner as well as an audio amplifier and speaker. Sound receiver with speaker.

Reflector

Lighting accessory helpful for spreading light as well as filling in shadows. Often made of lightweight reflective metal or poster board covered with metallic material.

Remote

[1] Videomaking performed "on location", outside controlled studio environment. [2] Equipment allowing from-a-distance control, usually without physical connections.

Rendering Time

The time it takes a DTV computer to compute a wipe or DVE from two video sources, typically 30 seconds to "render" a 30-frame effect.

Resolution

Amount of picture detail reproduced by a video system, influenced by a camera's pickup, lens, internal optics, recording medium, and playback monitor.

RF (Radio Frequency)

(Radio Frequency) Combination of audio and video signals coded as a channel number, necessary for radio /television broadcasts as well as some closed-circuit distribution.

RF Converter

Device that converts audio and video signals into a combined RF signal suitable for reception by a standard TV.

RGB

(Red, Green, Blue) Video signal transmission system that differentiates and processes all color information in separate red, green, and blue components—primary colors of light—for optimum image quality.

Roll

Text or graphics—usually credits—that move up or down the screen, typically from bottom to top. Produced with character generator.

Safe Title Area

The area on a monitor defined as 80 per cent of the screen area measured from the centre. Keeping the title within this area insures that the complete title will be visible on all TV sets.

Sampling Rate

The sampling rate for all Compact Disc recordings is 44.1 KHz.

Scan Converter

Device that changes scan rate of a video signal, possibly converting it from non-interlaced to interlaced mode. For example allows computer graphics to be displayed on a standard video screen.

Scan Line

Result of television's swift scanning process which sweeps out a series of horizontal lines from left to right, then down a bit and left to right again. Complete NTSC picture consists of 525 scan lines per frame.

Scan Rate

Number of times a screen is "redrawn" per second. Computer displays operate at different scan rates than standard video.

Scene

In the language of moving images, a sequence of related shots usually constituting action in one particular location.

Script

Text specifying content of a production or performance, used as a guide. May include character and setting profiles, production directives (audio, lighting, scenery, camera moves), as well as dialogue to be recited by talent.

Secam (Sequential Color and Memory)

(Sequential Color and Memory) 625 line 25 frame per second television signal standard used in France, former communist-block countries and a few other areas including parts of Africa.

Shot

All pictorial material recorded by a camera.

Shotgun

Highly directional microphone with long "barrel" designed to pick up sound from extreme subject-to-mike distances.

Signal-to-Noise Ratio

(S/N) provides relationship between signal strength and a medium's inherent noise. Video S/N indicates how grainy or snowy a picture will be, plus color accuracy; audio S/N specifies amount of background tape hiss present with low- or no-volume recordings. Higher the S/N the cleaner the playback.

SMPTE

Time-code standard for film, video, and audio named after the Society of Motion Picture and Television Engineers, which sanctions standards for recording systems in North America.

Snow

Electronic picture interference; resembles scattered snow on the television screen. Synonymous with chroma and luma noise.

Soft Light

Light which is diffused and creates very soft shadows.

Sound Effects

Contrived audio, usually pre-recorded, incorporated with a video soundtrack to resemble the real thing. Blowing on a microphone, for example, might simulate wind to accompany hurricane images.

Soundtrack

The audio portion of a video recording, often multi-faceted with voiceover, background music, sound effects, etc. is used as soundtrack.

Special Effects

Tricks and illusions—electronic or on camera—employed in film and video to define, distort, or defy reality.

Special Effects Generator

Video signal processor with vast, but varying, image manipulation capabilities involving patterns and placement as well as color and texture: mixing, multiplying, shrinking, strobing, wiping, dissolving, flipping, colorizing, etc.

Spotlight

Radiates a well-defined directional beam of light, casting hard, distinct shadows. Best used to focus illumination on individual subjects, whereas floodlights blanket broader areas.

Stereo

Sound emanating from two isolated sources, intended to simulate pattern of natural human hearing.

Still Store

Device which stores individual video frames, either in analog or digital form, allowing extremely fast access time.

Storyboard

Series of cartoon-like sketches illustrating key visual stages (shots, scenes) of planned production, accompanied by corresponding audio information.

Superimposition

Non-inherent titles or graphics appearing over an existing video picture, partially or completely hiding areas they cover. A picture superimposed on another can appear transparent.

Super-VHS

(S-VHS, S-VHS-C) Improved version of VHS and VHS-C videotape formats, characterized by separate carriers of chrominance and luminance information, yielding a sharper picture.

S-video

Also known as Y/C video, signal type employed with Hi 8 and S-VHS video formats. S-video transmits chrominance and luminance portions separately via multiple wires.

Sweetening

Audio post-production where audio is corrected and enhanced. Music, narration and sound effects are mixed with original sound elements.

Switcher

Switcher permits video signal mixing from two or more sources—cameras, time base correctors, character generators—for dissolves, wipes, and other clean transition effects.

Sync

(Synchronization) Horizontal and vertical timing signals or electronic pulses—component of composite signal, supplied separately in RGB systems. Aligns video origination (live camera, videotape) and reproduction (monitor or receiver) sources.

Synchronous Sound

Audio recorded with images. When the mouth moves, the words come out.

Talent

Generic term for the people or creatures assuming primary on-screen roles in a videotaping.

Take

An individual shot. When time and budgets permit, many takes may be filmed of the same shot.

Target Audience

The intended viewers.

Telecine Converter

Imaging device used in conjunction with a movie projector and camcorder to transfer film images to videotape.

Telephoto

Camera lens with long focal length, narrow horizontal field of view. Opposite to wide-angle, captures magnified, close-up images from considerable distance.

Teleprompter

Mechanical device that projects and advances text on mirror directly in front of camera's lens, allowing talent to read their lines while appearing to maintain eye contact with viewers.

Test Pattern

Various combinations of converging lines, alignment marks, and gray scales appearing on screen to aid in video equipment adjustment for picture alignment, registration, and contrast. Often viewed on broadcast television in off-air hours.

Three-point Lighting

Basic lighting approach employing key, back, and fill lights to illuminate subject with sense of depth and texture. Strategic placement imitates natural outdoor lighting environment, avoids flat lighting.

Three-quarter-inch

(U-matic) Most popular professional/industrial video format employing larger cassettes and three-quarter-inch tape, as opposed to the half-inch width of VHS and Beta "consumer" formats. Related equipment is generally larger and sturdier, format's recording considered superior. U-MATIC - trade name for the 3/4 inch video format developed by Sony. ¾ SP is an enhanced version.

Tilt

Camera "pan" move in a vertical direction, down or up, from a stationary position. Follows movement, contrasts differences in size between two subjects, or gives viewer point-of-view sense of a subject's height.

TBC (Time Base Corrector)

A device to correct timing errors which can cause unstable edits. These errors are caused by the slight mechanical defects inherent in the playback of videotape machines.

Tilt

Movement of the camera on its vertical axis.

Time-Code

A system of numbering each frame of video with a unique address divided into hours, minutes, seconds and frames. There are 30 video still frames per second.

Time Line Editing

A computer-based method of editing in which video and audio clips are represented on a computer screen by bars proportional to the length of the clip. These bars can be moved and re-sized along a grid whose horizontal axis relates to the time of the program.

Titling

Process or result of incorporating on-screen text as credits, captions, or any other alphanumeric communication to video viewers.

Tracking

[1] Lateral camera movement aligned with moving subject; background appears to move. [2] Positioning of video and/or audio heads over a videotape's recorded signals.

Tripod

Three-legged camera mount offering stability and camera placement/movement consistency.

Trucking Shot

A camera movement, which films the subject from side to side.

Tuner

Television and VCR component that receives RF signals from an antenna or other RF sources and decodes into separate audio and video signals.

Two-shot

Camera view including two subjects, generally applicable to interview situations.

Type C

SMPTE standard for 1 inch non-segmented helical video recording format.

Umbrella

Lighting accessory available in various sizes usually made of textured gold or silver fabric. Facilitates soft, shadowless illumination by reflecting light onto a scene.

Uni-directional

Highly selective microphone pickup pattern, which rejects sound coming from behind while absorbing that from in front.

Vectorscope

Electronic testing device that measures a video signal's chrominance performance, plotting qualities in a compass-like graphic display.

Vertical Interval

Indicates the vertical blanking period between each video field.

Contains additional scan lines above the active picture area into which non-picture information (captioning, copy protection and other control signals) may be embedded.

Vertical Sync

Synchronizing pulses used to define the end of one television field and the start of the next, occurring at a rate of approximately 59.94 Hz.

VITC

(Vertical Interval Time Code pronounced vitSEE) This type of time-code is recorded in the vertical blanking interval above the active picture area. Can be read from videotape in the "still mode."

VHS

(Video Home System) Pre-dominant half-inch videotape format developed by Matsushita and licensed by JVC.

VHS-C (VHS compact)

(VHS compact) Scaled-down version of VHS using miniature cassettes compatible with full-size VHS equipment through use of adapter.

Videocassette Recorder

(VCR) Multi-function machine intended primarily for recording and playback of videotape stored in cassettes.

Videographer

A video photographer.

Voiceover

Narration accompanying picture, heard above background sound or music, without the narrator being seen on camera. Talk's typically applied to edited visual during post-production.

Waveform Monitor

Specialized oscilloscope testing device providing a graphic display of a video signal's strength.

White Balance

A color camera function, which determines how much red, green and blue is required to produce a normal-looking white. Shots made with improper white balance will have an abnormal color tint.

Wide-angle

Camera lens with short focal length and broad horizontal field of view. Opposite of telephoto, supports viewer perspective and tends to reinforce perception of depth.

Windscreen

Sponge-like microphone shield, thwarts undesirable noise from wind and rapid mike movement.

Wipe

Picture transition from one scene to another wherein the new scene is revealed by a moving line or pattern. In simplest form, simulates a window-shade being drawn. More sophisticated variations include colorized wipes, quivering wipes, triangle wipes, and venetian blind wipes.

Wireless Microphone

Consisting of radio transmitter and receiver, utilizes low-power radio signal for cable-free operation.

Wow and Flutter

Sound distortions consisting of a slow rise and fall of pitch, caused by speed variations in audio/video playback system.

XLR

Three-pin plug for three-conductor "balanced" audio cable, employed with high-quality microphones, mixers, and other audio equipment. Also called "Cannon."

Y

Symbol for luminance, or brightness, portion of a video signal; the complete color video signal consists of R,G,B and Y.

Zoom

Variance of focal length, bringing subject into and out of closeup range. Lens capability permits change from wide-angle to telephoto, or vice versa, in one continuous move. "Zoom in" and "zoom out" are common terms.

Zoom Ratio

Range of a lens' focal length, from most "zoomed in" field of view to most "zoomed out." Expressed as ratio: 6:1, for example, implies same lens from same distance can make same image appear six-times closer.